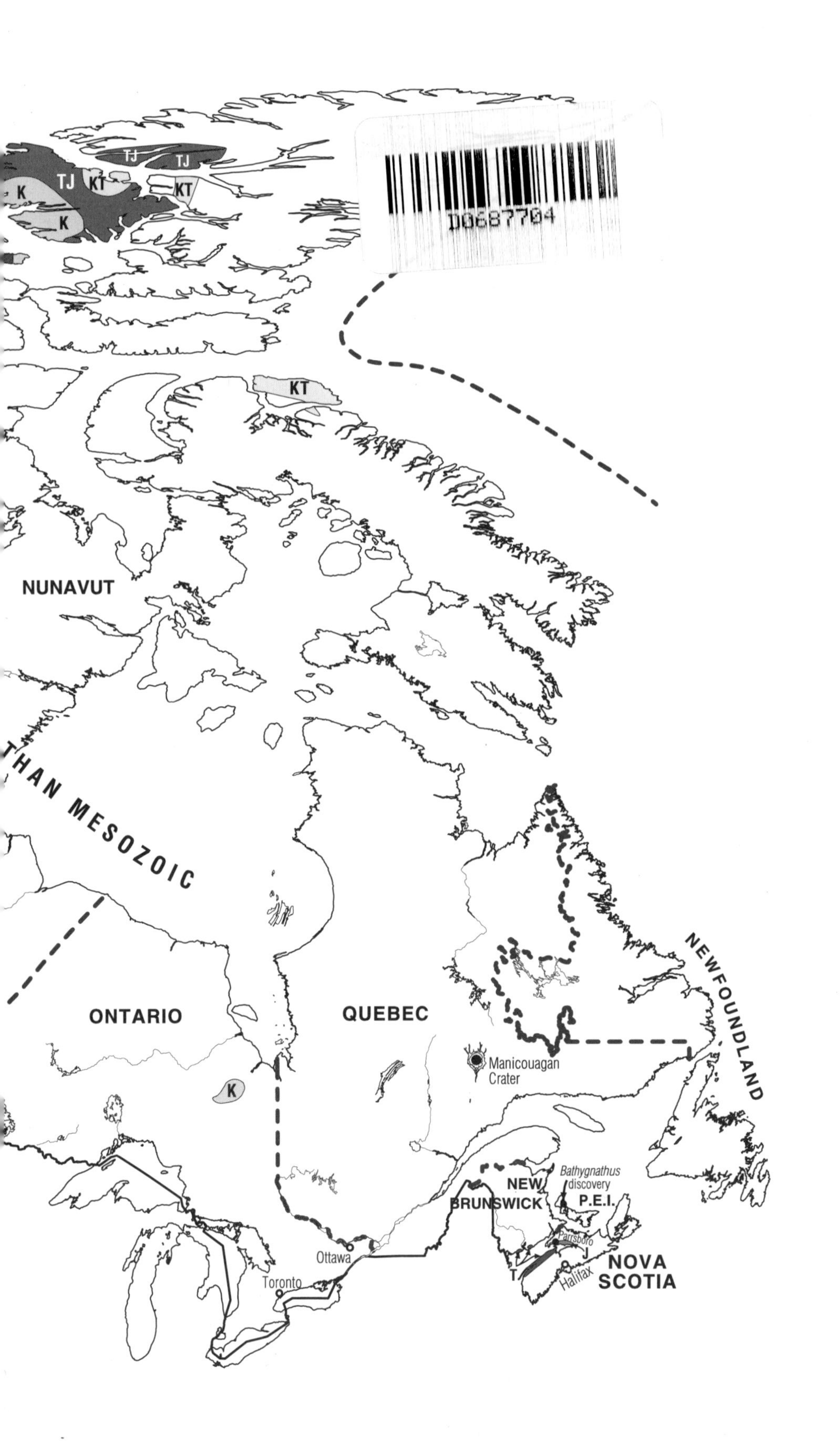
TJ
TJ
TJ
KT
KT
K
K
KT
NUNAVUT
THAN MESOZOIC
ONTARIO
QUEBEC
K
Manicouagan
Crater
NEWFOUNDLAND
NEW
BRUNSWICK
Bathygnathus
discovery
P.E.I.
Parrsboro
J
T
Halifax
NOVA
SCOTIA
Ottawa
Toronto

INTO THE DINOSAURS' GRAVEYARD

Canadian Digs and Discoveries

DAVID SPALDING

Foreword by Dr. Philip Currie

Doubleday Canada

Canadian Cataloguing in Publication Data

Spalding, David A.E., 1937-
Into the dinosaurs' graveyard: Canadian digs and discoveries

Includes index
ISBN 0-385-25762-7

1. Dinosaurs – Canada. I. Title.

QE862.D5S6537 1999 567.9'0971 C99-930986-2

Jacket photograph courtesy Comstock/Bob Rose
Inset photograph Neg. No. 19493 (Photo. by Brown) courtesy Dept. of Library Services, American Museum of Natural History
Jacket and text design by Heather Hodgins
Endpaper design by Susan Thomas/Digital Zone
Printed and bound in the USA

Published in Canada by
Doubleday Canada, a division of
Random House of Canada Ltd.
105 Bond Street
Toronto, Ontario
M5B 1Y3

Care has been taken to trace ownership of archival material in this book. The publishers will gladly receive any information that will enable them to rectify errors or omissions affecting references or credit lines in subsequent printings.

BVG 10 9 8 7 6 5 4 3 2 1

A Dedication

To Bill Sarjeant

in celebration of more than forty years
of collaboration and friendship

and

A Curse

For the unknown civil servant
who kept the card index to correspondence
of the Geological Survey of Canada

. . . but threw away the letters!

CONTENTS

ACKNOWLEDGEMENTS

Researching this story has been an enjoyable part-time project over some twenty years, and it would be difficult to provide proper acknowledgement to everyone who helped me during that period. Financial assistance for related projects came from Alberta Culture and Alberta Literary Arts Foundation, and information and encouragements came from many colleagues at the Provincial Museum of Alberta, and other museum professionals and dinosaur enthusiasts I have met over the years. Parts of the story have been published in other forms as articles in *Alberta Studies in the Arts and Sciences*, *NeWest Review*, and *Vertebrate Fossils and the Evolution of Scientific Concepts*; in my edition of Sternberg's *Hunting Dinosaurs* (NeWest Press); and in my book *Dinosaur Hunters* (Key Porter).

In singling out those who provided assistance during the run-up to takeoff, I apologize to many others who are not specifically mentioned. Al Purdy gave me permission take my title from his poem. Notable assistance was provided by Dale Russell and colleagues at the Canadian Museum of Nature; Jim Ebbels and Linda Strong-Watson of the Ex Terra Foundation; Ken Adams and Tim Fedak at the Fundy Geological Museum; Robert Carroll at the Redpath Museum, Montreal; the late Loris Russell, Chris McGowan, Brian Noble, and Hans-Dieter Sues at the Royal Ontario Museum; Tim Tokaryk at the Royal Saskatchewan Museum of Natural History; Dennis Braman, Don Brinkman, Philip Currie, Dave Eberth, Bruce Naylor, Betsy Nicholls, Monty Reid, Wendy Sloboda, Darren Tanke, and their colleagues at the Royal Tyrrell Museum; and Bill Sarjeant at the University of Saskatchewan.

Many friends extended their hospitality while I researched this book, providing accommodation and meals and sometimes transportation. These particularly include (from west to east) George Melnyk and Julia Berry in Calgary; David and Elna Nash in Edmonton; Monty and Pat Reid in Drumheller; Bill and Peggy Sarjeant in Saskatoon; Bruce Nash and Nona Arneson in Toronto; and Rob Keip and Patricia Morley in Ottawa.

Deb Lister and other staff at the Royal Tyrrell Museum library; archivists at the American Museum of Natural History, British Columbia Archives and Record Service, Canadian Museum of Nature,

Glenbow-Alberta Institute, and National Archives of Canada; Sandra Chapman at the Natural History Museum (London); and Gertrude Bloor McLaren at the University of Alberta were particularly helpful, seeking out documents and photographs and giving permission to quote from their holdings.

Greg Hall and Jock Mardres helped to keep my computer functioning efficiently. My brother-in-law Gavin Clarke provided additional computer support.

Ed Andrusiak, Mary-Anne Neal, Sher O'Hara, and Sharon Sinclair are among many friends and associates who passed me stories, cuttings, Internet addresses, and other useful information.

Susan Mayse not only connected me with Doubleday Canada but also was a pillar of support in editing the manuscript; Susan Broadhurst undertook a thorough copyedit; while John Pearce and Kathryn Exner provided enthusiastic support.

Bill Sarjeant (who is specially acknowledged in the dedication) helped me in ways too numerous to particularize, and my wife, Andrea, made the book possible.

PHOTO SOURCES

Glenbow Archives: pages 27, 76, 105

National Archives of Canada: page 5

W.A.S. Sargeant: pages 91, 123, 132, 143, 147, 163

David A.E. Spalding: pages 3, 29, 97, 107, 128, 134, 141, 145, 155, 161, 168, 171, 177, 181, 186, 196, 197, 216, 222, 241, 242, 257, 258, 260, 267, 273, 274, 280

University of Alberta Archives: page 118

into the dinosaurs' graveyard
where a movement or an action
begun in the past
is never completed
but continues now

"Lost in the Badlands", Al Purdy

FOREWORD

In the twentieth century, the study of dinosaurs has gained a loftier position than many of the sciences. It is a study as serious and rigorous as physics or chemistry, yet it maintains a hold on the general public that can only be referred to as a cultural phenomenon. From *Jurassic Park,* and a purple caricature named Barney, to the almost certain resolution of the mystery of bird origins, dinosaurs have invaded our lives. Palaeontologists have evolved from simply trying to justify why they study dinosaurs to supporting a multi-billion dollar dinosaur industry that capitalizes on their appeal through books — both scientific and fictionalized — amusement parks, feature films, cartoons, comics, museum displays, television programs, toys, and countless other projects. Yes, interest in dinosaurs is no longer restricted to a fringe group of scientists; there is something about these magnificent creatures that has struck deep into the very heart of our existence.

My love affair with dinosaurs began when I was six years old, and grew into a profession in 1976 when I was lucky enough to be hired as a dinosaur palaeontologist at the Provincial Museum of Alberta in Edmonton. I was still a graduate student and was competing with more qualified applicants for the position. But David Spalding, who at the time was in charge of the Natural History section, agreed to give me a chance — a decision that has certainly helped my career. I never grew out of my obsession with dinosaurs. They have become my life, appealing to me from so many angles that I can easily enjoy the work aspect of dinosaurs as well as the fun side of watching movies or reading books about them. On many occasions, Dave and I talked about our common interests in dinosaur biology, the history of dinosaur hunting, and dinosaurs in fiction. Those were good days at the Provincial Museum, and as a supervisor, Dave gave me the freedom to develop a program that attracted international attention, playing a significant role in the "dinosaur renaissance."

Over the years, David Spalding has supplemented his experience in dinosaur palaeontology by probing deep into the archives of many of the finest museums in the world for information on the collection of dinosaurs. I doubt there is anyone else in the world who knows as much about this subject. The stories Dave discovered — and sometimes participated in — are as fascinating as the dinosaurs themselves, and he has assumed an important role in making them available to the public. As you enjoy this book, be aware that it is also a primary source of information documenting the history of a science in the making.

Dr. Philip J. Currie
Curator of Dinosaurs
Royal Tyrrell Museum of Palaeontology
Drumheller, Alberta

1

BURIED TREASURE

"Just look at Arrowsmith's little map of British North America. . . . You will see that Canada comprises but a small part of it. Then examine the great rivers and lakes which water the interior between . . . Hudson's Bay, and the Pacific Ocean. . . . It will become a great country hereafter. But who knows anything of its geology? . . . How insignificant would be the expense . . . in comparison with the advantage which might result."

WILLIAM LOGAN, 1845

DEATH IN THE AFTERNOON

On a sunny afternoon, the dinosaur lumbered across a sand dune, in company with a few others. When it paused to gaze around, it saw deep valleys bottomed by lakes, active volcanoes, and scattered forests. It slithered down the side of the dune and sought shelter in a sparse clump of cycads. After a pause, it began to tear at the fern-like leaves a few inches from its face, angling its long neck to get closer to fresh forage. At about 2.5 metres (seven feet) long, it was close to human size, but no humans would be around for comparison for another 200 million years.

The dinosaur looked somewhat like a skinny crocodile, except that its legs did not angle away from its body, but supported it like slender pillars. As its jaws searched higher and higher, it rested its long tail on the ground and raised its body on its long hind legs. One short foreleg rested against the tree trunk while the other groped for the higher leaf stems now within its reach, and pulled them down to its mechanically chomping mouth.

After ripping off the leaves within reach, the dinosaur clumsily dropped to all fours and ambled into a grove of primitive conifers, where it resumed eating. Gradually it worked from plant to plant until it was under the shade of a lava cliff, eroding into a coarse boulder slope. It pulled too hard at a stem that was holding up a litter of boulders, and with a sudden rumble, part of the cliff collapsed. A squeal, then the dinosaur lay, seriously injured, its hindquarters partly buried beneath the tumble of rocks. As it died, a pack of small predatory land crocodiles found the body, and tore at the parts they could reach. The head and neck vertebrae were carried away and eroded, while the parts of the body under the rock fall decayed undisturbed. Over the weeks and months that followed, the wind blew the sand dune forward until it covered the base of the cliff, the boulders, and the remains of the dinosaur.

THE REAL JURASSIC PARK

Over thousands of centuries the rocks containing the dinosaur's bones were buried by other sediments, which were squeezed, tilted, and lifted by moving continental plates. More recently, the forces of erosion have exposed those rocks, which now appear in a line of cliffs, running west from Wasson Bluff, near Parrsboro in western Nova Scotia. On a recent June day I walked along the beach, far from my home on Canada's western seashore. In the thirty years I had been interested in Canada's dinosaurs, I had never been to this spot — now known to be as important a site as any in the country.

On my left the world's biggest tides ebbed out of the Minas Basin into the Bay of Fundy, dropping more than fourteen metres (forty feet) and gradually revealing miles of sand and mud. However, my attention was turned towards the cliffs, where the evidence of a much more ancient landscape is revealed in fascinating detail. "This is the real Jurassic Park," says Ken Adams, director of the nearby Fundy Geological Museum.

Sandstones, shales, and gravels show old lake beds, while lava beds indicate flows from long-dormant volcanoes. Further along, a notch cut in a lava bed shows erosion by an ancient lake, whose wind-driven waves had worked on ancient cliff in the same way that the sea does today. Walking west, I came to reddish sandstones, whose cross-bedding shows that they are the remains of old dunes. At intervals, lava boulders lie in

Fundy's cliffs yield early dinosaurs.

loose layers in the sandstone, showing how they fell from the ancient cliffs into the sand. As I approached another buried cliff, I saw the remains of the talus slope of lava boulders that concealed the dinosaur whose story opened this chapter. Other dinosaur skeletons have been found here, some of which have not yet been excavated. The buried landscape, and the other life forms found in the rocks, yield the clues for my opening narrative.

Science doesn't yet have a certain name for this dinosaur — the loss of its skull makes it difficult to compare these scanty bones with more complete dinosaurs found elsewhere. But we do know it belongs to the group known as the prosauropods; perhaps ancestors of the great long-necked and long-tailed sauropods that are among the best-known dinosaurs. To the palaeontologist — the student of fossils — even these fragmentary finds are buried treasure, yielding information more precious than gold about the remote past of the earth, and the slow procession of life that preceded us.

The Fundy prosauropods do not look like rulers of the earth. Their skeletons are incomplete, and their bones are small, sometimes fragmentary, and hard to find and separate from the matrix of hard rocks. It is not surprising that these early dinosaurs from eastern Canada were barely known to scientists a decade or so ago, and are only now becoming visible to the residents of the Maritimes and their visitors. For most residents of this huge country — if they are aware that Canada has dinosaurs at all — Canadian dinosaur finds call to mind the prairies and badlands of the west, and the much larger dinosaurs they have produced. Yet the Maritimes not only have the oldest dinosaurs in the country, but also the story of the discovery of Canadian dinosaurs, which starts — somewhat obliquely — in the eastern provinces, more than 140 years before the discovery of the "Jurassic Park" prosauropod.

A DISTINCT TRIBE OF SAURIAN REPTILES

John William Dawson was born a mere 130 kilometres (80 miles) from Parrsboro, in Pictou, Nova Scotia, in 1820. His Scottish Presbyterian father ran a bookshop, and young Dawson read widely of his stock, which included both theology and science. Dawson attended Pictou Academy, began collecting fossils, and in 1840 was sent on the long Atlantic voyage to study briefly at Edinburgh University. When he landed in England, he was congratulated on the quality of his English.

Edinburgh was one of the few universities at that time to offer any teaching in geology, but Dawson studied under Robert Jameson, last bastion of eighteenth-century ideas. These were soon to give way to modern geological thinking, which had been first popularized by John Playfair (1748–1819), Professor of Mathematics at the same university, almost four decades earlier. Dawson found that Scottish geology had some similarities to that of Nova Scotia, and on his return in 1841 he resumed his explorations of the local geology.

Playfair's geological ideas were being extended by another Scot, Charles Lyell (1797–1875), whose three-volume *Principles of Geology* had appeared during the 1830s. Lyell had studied at Oxford under William Buckland (1784–1856), who described the first carnivorous dinosaur in 1824. The first herbivorous dinosaur was described by Sussex doctor Gideon Mantell (1790–1852) in the following year. Britain's leading anatomist, Richard Owen (1804–1892), recognized similarities between these distinct creatures in an 1841 paper on British fossil reptiles presented in Plymouth. When it was published in the following year, Owen recognized "a distinct tribe or suborder of saurian reptiles for which I would propose the name Dinosauria."

Lyell made his first trip to North America in 1841, for he had been invited to present his new ideas to the élite of American intellectuals through a course of lectures at the Lowell Institute in Boston. Coal was

J. W. Dawson, pioneer palaeontologist.

the fuel of the developing Industrial Revolution, and new information came from every mine, so Lyell planned to compare the coal-bearing rocks on both continents. He brought letters of introduction to English mining engineers working in New Brunswick, through whom he was introduced to Dawson. In later life Dawson remembered how Lyell examined his fossil collection, and "listened with interest to what I could tell him of the geology of the beds in which they occurred." They went in the field together, and Dawson sought Lyell's advice on furthering his geological career.

After resuming his studies in Edinburgh in 1846, Dawson returned to Nova Scotia the following year with a wife, Margaret Mercer. Though Lyell had "cautioned [Dawson] against entering into educational work, unless of such a kind as to give time for research," there were few other opportunities open to an aspiring scientist, and he turned to teaching to support his family. He became in rapid succession a professor at Dalhousie College, then Superintendent of Education for Nova Scotia. His travels in the province allowed him to extend his geological researches, and in 1855 he published his pioneering regional study, *Acadian Geology*.

OF GREAT INTEREST TO NATURALISTS

Lyell and Dawson found the world's first reptile remains in coal measures at Joggins, in rocks 100 million years older than those at Parrsboro. However, in all his travels in Nova Scotia, Dawson did not find any dinosaurs. He became interested in the red sandstones, which were also found on Prince Edward Island, where he went to see them. Around 1845, a landowner near New London, Donald McLeod, had dug a well into the red sandstones, and the diggers found an unexpected treasure — a spectacular bone, bearing seven fierce-looking teeth. McLeod invited Dawson to buy the specimen. Although Dawson could not afford to purchase it himself, he knew "it would be of great interest to naturalists, if examined and described by a competent anatomist," and so offered to negotiate its sale.

In later years Dawson remembered his early difficulties, when there " . . . were no collections in geology or natural history except the private cabinets of a few zealous workers . . . no special schools of practical science, no scientific publications, and scarcely any printed information accessible." Dawson must have looked in vain for a suitable repository in his neighbourhood for such a spectacular fossil. Although a few fledgling institutions existed, none had funds or "a competent anatomist."

Dawson consulted Lyell during his 1852 North American visit. At this time a fossil found in a British colony normally would have been sent to Richard Owen (who had described other vertebrate fossils found by Dawson), but the two had fallen out when Lyell backed a rival candidate for a position Owen wanted for himself. Instead, Lyell drew the specimen to the attention of Joseph Leidy, of the Academy of Natural Sciences in Philadelphia.

THIS DEEP-JAWED MONSTER

Unlike Canada, the neighbouring United States had institutions ready and willing to buy, house, and describe a spectacular specimen. The Academy of Natural Sciences in Philadelphia had existed since 1812, and by 1848 had important collections of fossils.

The fossil collections were in the care of Joseph Leidy (1823–1891), who had qualified as a doctor, thus obtaining the only anatomical training available in the early nineteenth century. He became Professor of Anatomy at the University of Pennsylvania, and then in 1854 Professor of Natural History at Swarthmore College. He was closely involved with the Academy, donating and describing items in the collections, and later serving as trustee and president. In his thirties, he was already describing vertebrate fossils, and although still inexperienced was well on his way to earning his later reputation as the father of American vertebrate palaeontology.

With the help of two friends, Leidy purchased the specimen for $30, and in 1855 he described it in the Academy Proceedings under the name of *Bathygnathus borealis* ("northern deep jaw"). Leidy felt it was a "new and remarkable genus" and underlined its importance by illustrating his paper with a coloured plate. He was wrong about the anatomy, for he described it as a lower jaw, when (as Owen gleefully pointed out) it was actually an upper. He was also cautious about its relationships, describing it merely as a "carnivorous lacertilian" or lizard. However, he did compare it with the first known carnivorous dinosaur, Buckland's *Megalosaurus.*

SCIENTIFIC PUBLICATION

Bathygnathus was a new discovery — a fossil of a kind of animal not known anywhere else in the world. Leidy was able to compare *Bathygnathus* to *Megalosaurus* because published information was available about the English dinosaur, and he in turn published a description and illustrations in the *Proceedings of the Academy of Natural Sciences, Philadelphia.*

The specimen on which a new species is based becomes known as the type specimen. Such specimens are preserved in public collections, and are the ultimate court of appeal if there is future confusion. Every new species of dinosaur enters the body of scientific knowledge in this way, when its name is published, along with a description (and, preferably, pictures) of its remains. During the nineteenth century, a number of periodicals and series publications became established in which new species were often described. These were published by museums or societies, surveys or universities, and had the advantage that they could be published quite quickly. The publications usually contained several papers on different topics, and were purchased by libraries of institutions working in the same field.

Palaeontologists (then and now) often circulate copies of their papers ("offprints") directly to other specialists in the same field, so that the information is shared as soon as possible. This is a courtesy to colleagues, but it also ensures that your priority is quickly established on any new species that you have named. The one thing every scientist dreads, when he is three-quarters of the way through preparing an account of a new species, is that someone else will have found similar specimens and will publish their findings first. Consequently, early descriptions were often very perfunctory — sometimes even sent in from the field by telegraph — causing endless problems in interpretation of the species in later years.

Nowadays, a palaeontologist may take years to prepare a thorough and detailed description of a new species of dinosaur, and it is likely to be published in detail. Still, occasionally two people publish the same species under different names, or (equally inconvenient) use the name you had planned for a quite different animal.

Leidy was a mentor for his younger follower, Edward Cope (1840–1897), who in 1867 was the first to specifically refer to *Bathygnathus* as a dinosaur. In ensuing decades Leidy himself, and Cope's rival Othniel Marsh (1831–1899), made it clear that they agreed with this interpretation. Edward Hitchcock (1793–1864) collected abundant fossil tracks from New England, and died convinced that they were made by birds. Leidy was not so sure. "Was this animal not one of the bipeds," he speculated of *Bathygnathus,* "which made the so-called bird tracks of the New Red Sandstone of the valley of the Connecticut?" As late as 1903 Dawson himself regarded *Bathygnathus* as a dinosaur, and Lawrence Lambe listed it as "the only known dinosaur from eastern Canada" in 1905.

Prince Edward Island naturalist Francis Bain (1842–1894) had a field day with the wonderful fossil, which he called "Monster Dinasurous." In purple prose, he pictures it in life: "A very Tamerlane in blood-thirst and cruelty must this reptile have walked forth in the history of the past. Even the scaled crocodiles and megalosaurians must have found their thick armature but poor defense against the onset of this deep-jawed monster."

Dawson, accepting that the fossil was a dinosaur, used it to date Prince Edward Island's red sandstones as Triassic. By the late nineteenth century, however, another group of fossil reptiles, the great sail-backed pelycosaurs, had been recognized in older Permian rocks in Texas. By

1905, comparison with these more complete fossils made it clear that *Bathygnathus* was not a dinosaur, but a pelycosaur. Moreover, ongoing study of the rocks in which it was found showed that they were older than any dinosaur-bearing beds. The findings of science are always subject to revision when new data comes along, and *Bathygnathus* — regarded for half a century as Canada's first dinosaur — has become an almost-forgotten footnote in Canada's dinosaur history.

Wann Langston of the National Museum of Canada made a study of *Bathygnathus* in the 1960s, when he borrowed the specimen and made a number of casts, one of which was destined for Prince Edward Island. In the absence of a provincial natural history museum, however, the cast was housed in various temporary quarters — even for a while in neighbouring New Brunswick. In 1992 it was placed in the University of Prince Edward Island's collections, where it is displayed from time to time.

Although widely regarded as Canada's first dinosaur, *Bathygnathus* was found before Prince Edward Island ceased to be an independent

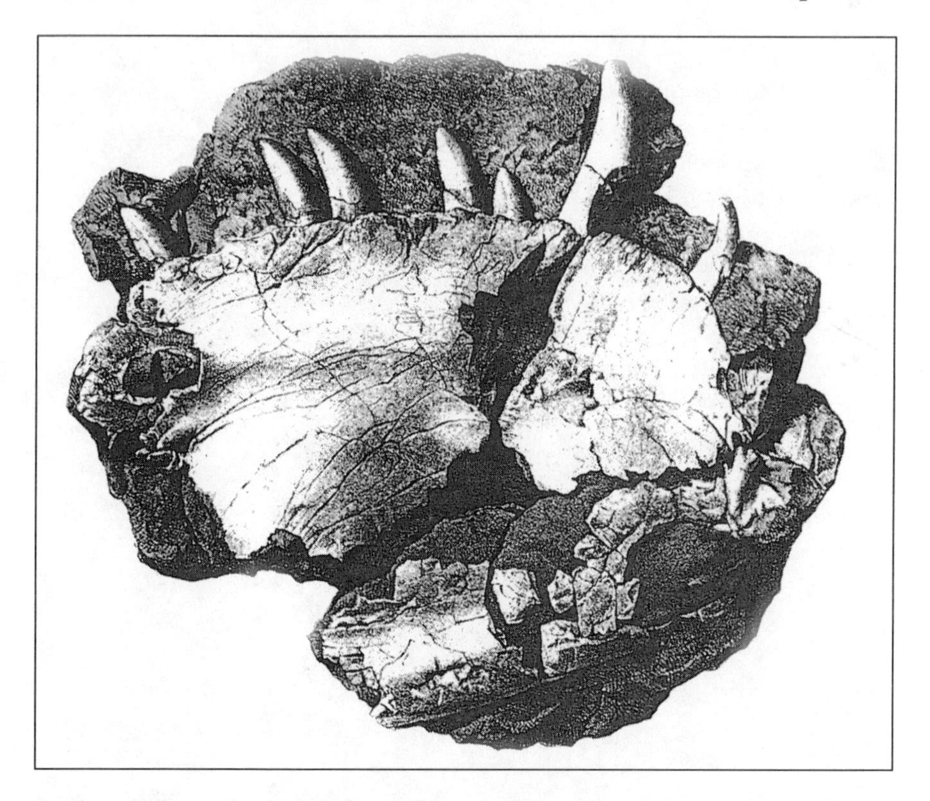

Bathygnathus, *Canada's first "dinosaur."*

province in 1873, and the fossil passed through Nova Scotia to the United States before 1867, when that province joined Ontario, Quebec, and New Brunswick to become the Dominion of Canada. Consequently (until Langston brought the specimen to Ottawa to make his casts) Canada's first dinosaur had never actually been on Canadian soil. Absurdly, the same is also true of the country's second "dinosaur."

TORONTO IS NOT VERY MUCH OUT OF THE WAY

*"Thomas Henry Huxley Esq. F.R.S.
Professor of Natural History in the
University of Toronto."*

THOMAS HENRY HUXLEY

In 1851, the English naturalist Huxley (1825–85) tried out a possible new title in a note to his fiancée. Huxley was in London, working on his collections from a round-the-world voyage as surgeon on the naval vessel *Rattlesnake.* He had met and become engaged to Henrietta Heathorn in distant Australia; only a paying position would allow them to marry and be together. Yet there seemed to be no suitable positions available in Britain, and an Australian lead came to nothing.

Lyell anticipated that Huxley (who had just been elected a Fellow of the Royal Society at age twenty-six) "will soon rank as one of the first naturalists we have ever produced." Clearly, he would be a prime candidate for a vacancy at the University of Toronto.

> Toronto is not very much out of the way, and the pay is decent and would enable me to devote myself wholly to my favourite pursuits. . . . it would answer my purpose very well for some years at that rate. . . . I think I shall have a very good chance of being elected; but I am told that these matters are often determined by petty intrigues.

Huxley's application included "twelve or fourteen testimonials from the first men," including Charles Darwin and Richard Owen. By May 1852 Huxley was complaining, "I have heard nothing of

Toronto, and I begin to think the whole affair, University and all, is a myth." There were rumours that "the chair will be given to a brother of one of the members of the Canadian ministry, who is, I hear a candidate. Such a qualification as that is, of course, better than all the testimonials in the world." Huxley went to a Belfast conference in order to lobby for the position, but became so desperate he considered giving up science altogether. "I can get honour in science, but it doesn't pay. . . . I may give up the farce altogether . . . and take to [medical] practice in Australia."

Late in 1853, the Rev. William Hincks (1794–1871) was appointed to the University of Toronto. His brother was Sir Francis Hincks (1807–1885), active in the Canadian government during this period. Rev. Hincks had studied for the ministry, and had taught mathematics and philosophy in England and natural history in Ireland. He was fifty-nine, and "intellectually quite rigid." In Toronto he primarily taught botany, where he presented an already outdated taxonomy, and his students remembered his teaching as "leaving much to be desired."

Soon after, the tide turned for Huxley. He received the Royal Society's gold medal and was appointed Professor of Palaeontology, and then Professor of Natural History at the Government School of Mines in London. It didn't pay as well as Toronto would have done, but "I am chief of my own department, and my position is considered a very good one." Huxley's restless mind followed up all the new discoveries in his science, and he described a fossil marine reptile in 1853. By 1859 he began a series of papers on vertebrate fossils, including thirteen dinosaur papers between 1865 and 1876. These included work on the relationship of birds and dinosaurs, the importance of which was fully understood only recently.

Huxley surely would have taken up these new fields in Toronto, so there could have been a brilliant vertebrate palaeontologist based in Canada soon after mid-century. As a friend of Lyell's, he surely would have been given responsibility for *Bathygnathus*, and certainly would have been consulted on the later western Canadian finds. The course of dinosaur studies in Canada might have been very different. But even in England, Huxley was not without influence on Canadian science, for he helped to train the man who did, eventually, discover Canada's first dinosaurs.

A FOSSIL SAURIAN FROM THE ARCTIC REGIONS

Until 1880, the Arctic Islands were not part of Canada, but were regarded as a British responsibility. When Sir John Franklin got lost in 1845 looking for the fabled Northwest Passage, many exploring expeditions were sent out to the barren wilderness of rocks and ice. In the summer of 1853, British naval officer Lieutenant Sherard Osborn (1822–1875) was leading a sledging party of British sailors along the north coast of Bathurst Island. Although he was searching for traces of Franklin's expedition, he also kept his eyes open for anything else of interest. When Osborn checked a rocky outcrop 46 metres (150 feet) up Rendezvous Hill for potentially useful minerals, he found and collected a number of fossil vertebrae.

Sherard Osborn was described as "sturdy in build with a square head, an aggressive nose, and upstanding hair." He had already had one tour of Arctic duty as lieutenant in command of a small screw steamer, when he was sent back again under the command of the unpopular Captain Belcher. His fossil vertebrae were stowed below when the ships became icebound. During the hard winter that followed, tempers frayed, and Osborn was arrested by his captain. When the expedition returned to London in 1854, Osborn asked for a court martial, but the Admiralty refused his request. They expressed their opinion of both parties by promoting Osborn, and in due course he became a Rear Admiral.

After three years of such dramatic events, it was unlikely that the fossils Osborn had collected remained high on his agenda. He perhaps gave individual vertebrae to interested parties, but only one of them has survived. This was presented to a Mr. Salter of the Geological Society, who in turn showed it to the Rev. Dr. Samuel Haughton, at this time Professor of Geology at Dublin University. Haughton realized he had a crushed, incomplete neck vertebra. He identified it as *Teleosaurus*, now known to be a Jurassic crocodile.

Salter gave the specimen to the director of the Natural History Museum, Dublin, who passed it to one of his staff, Andrew Leith Adams (1827–1882). Adams had been a military surgeon who had travelled widely and written about the natural history of several colonies, including Canada. After seven years in India he had developed a particular interest in elephants, fossil and modern, and on his retirement from the army had joined the staff of the Royal College of Science in Dublin. Adams agreed with Haughton that the fossil was reptilian, but in 1877 he described it under the name of *Arctosaurus osborni*

(Osborn's arctic reptile). However, he felt there was insufficient data to "establish reliable comparisons between it and the fossil genera of the Mesozoic formations."

At the Natural History Museum, Richard Lydekker (1849–1915) regarded *Arctosaurus* as a Triassic carnivorous dinosaur, and for years (as no new material was found) it appeared on the lists of dinosaurs with no further comment, being classified variously as a prosauropod or a carnivorous dinosaur (theropod). More recently it has been suggested that it may be a turtle, and in 1989 the National Museum of Canada's Dale Russell assigned it to an obscure group of Triassic reptiles known as the trilophosaurs. Nevertheless, it was still being mentioned as a possible dinosaur as recently as 1997.

TREASURE IN THE EARTH

In the early years of the nineteenth century, the geological research that took place in the British provinces of North America was at first by-product of a variety of expeditions and amateur efforts by military and naval officers. Soon, local, though unconnected, initiatives were taking place. An English geologist was brought out to look at Newfoundland, while Dr. Abraham Gesner, a medical man based at Parrsboro, was hired to do a geological survey of New Brunswick.

In 1842 William Logan (1798–1875) was hired to undertake a geological survey of the Province of Canada, a region stretching through the southern parts of today's provinces of Ontario and Quebec. Logan was born in Montreal, and studied at Edinburgh (though his subjects, apparently, did not include geology). He had undertaken important researches in Britain and was recommended for the survey by a number of leading British geologists, including Buckland and de la Beche, the director of the British Geological Survey. Logan's ambitions were not limited to Upper Canada, and in his letter of 1845 (quoted at the head of this chapter) he was angling for the chance to explore the west in the employ of the British Survey. With Confederation in 1867, Logan's limited mandate grew substantially, and in 1870 the prairies of the far interior were placed in Canada's care. But by this time, Logan had handed over the reins to a successor.

The expansion and exploration of the territory that is now Canada was fuelled by the search for wealth. Starting with fisheries, whaling, and the fur trade, the country continued to grow through logging and agriculture. The Survey was particularly important in the search for buried treasure — resources of coal, and later a wealth of metal ores and other fossil fuels. Uneconomic resources such as fossils were only sought incidentally. Fossils were known to be useful for interpreting the structure of the rocks, and a few people were interested in their wider scientific significance, but until Darwin produced a credible basis for evolutionary ideas, they were regarded as trinkets rather than as treasures. Logan's view of the advantages of a geological survey of the west did not include dinosaurs.

DINOSAUR DEPOSITS IN CANADA

During the century and a half of the Survey, Canada's geology has become better understood. A simple outline — a gross simplification of a very complex geology — will make it easier to understand the discoveries of dinosaurs in this huge country.

The Canadian Shield, the ancient core of North America, occupies most of Ontario, Quebec, Labrador, and Newfoundland, as well as parts of Manitoba, Saskatchewan, Alberta, and Nunavut. Its rocks are Precambrian, all more than 570 million years old. The rocks surrounding the Shield have been added to it through time, so that the oldest are generally closest to the Shield, and the youngest are further away. The mountain chains between the Rockies and the Pacific Ocean are made up of rocks that were formed elsewhere on the earth's surface and picked up by the North American continent as it drifted westward. They contain rocks of all periods and great diversity.

Large areas of the Maritime provinces, Gaspé and southern Ontario, southcentral Manitoba and parts of northeastern Alberta, are situated on rocks from the next era, the Paleozoic ("ancient life"). During this period, fish, amphibians, and the first primitive reptiles appeared, and some examples of all of these have been found in eastern Canada. Parts of Gaspé and Prince Edward Island — the source of *Bathygnathus* — are now known to be made of rocks of the Carboniferous and Permian periods, the latest Paleozoic periods. The most southerly large cities of Ontario and Quebec — Toronto, Ottawa, and Montreal — are all founded on these rocks.

Dinosaurs only appeared in the next era, the Mesozoic ("middle life"). This has been divided by geologists into three great periods: Triassic (245–208 m.y.), Jurassic (208–146 m.y.), and Cretaceous (146–65 m.y.). The abbreviation "m.y." is convenient for "million years (ago)."

The Triassic has limited distribution in Canada, notably in British Columbia and a pocket in Nova Scotia. Some of the richest Triassic fossil deposits in the world are found at Wapiti Lake in northern British Columbia. They are marine, and contain fish, swimming reptiles, and even a terrestrial gliding reptile — but they are too old to include dinosaurs. Later Triassic rocks are now known from the west coast of Nova Scotia, and have produced the oldest dinosaurs in Canada — perhaps the oldest in North America.

Nearby are the Lower (older) Jurassic rocks I visited at Parrsboro. Other Lower Jurassic deposits are in the northern Rockies. Canada has no equivalent of the Upper (later) Jurassic Morrison beds of the western U.S., which contain abundant dinosaur remains. Higher Jurassic rocks (meaning those found at higher levels of the geological column) are found in the southern Canadian Rockies, where dinosaur footprints have been found in the uppermost beds.

Canada's major dinosaur discoveries are from the Cretaceous period. Some Lower Cretaceous beds are found in northern and eastern British Columbia, where they are now known to contain abundant footprints. Others are in northern Alberta and central Saskatchewan, but are generally concealed by forest. The richest in fossils are the Upper Cretaceous beds that underlie most of the western plains. Remains of land animals such as dinosaurs are found mainly in sediments formed in rivers and lakes, and we are fortunate that most of these western beds are of this kind. However, there was a period during the Upper Cretaceous when a shallow sea expanded across the prairie region, and the rocks deposited in this sea contain remains of marine reptiles and fish instead of dinosaurs. Marine rocks of the same age occur in the Nanaimo Basin, mainly on the east side of Vancouver Island, and have also produced some marine reptiles, a single dinosaur tooth, and some bird bones.

After the Cretaceous period, the Cenozoic Era began, and (other than the birds) dinosaurs were extinct. Some Cenozoic deposits lie on top of Cretaceous rocks on parts of the plains, and occur in pockets elsewhere. Some of these have fossil birds and mammals, which also appear in deposits left after the last part of the Cenozoic Era, in the Pleistocene. In Canada, the Pleistocene period is represented mainly by deposits associated with the great glaciations, which lie on top of any older rocks.

The Arctic Islands generally have a similar geological structure. The Canadian Shield extends through most of Baffin Island in the east, then a broad belt of Paleozoic rocks makes up the central islands. Mesozoic rocks, mainly Cretaceous, occur in a belt along the western islands, from Banks up to Ellesmere. Here, too, real dinosaurs have been found in recent years.

GEOLOGICAL STRATA IN CANADA

Geologists summarize the sequence of the rocks in table form, with the youngest at the top. Dates indicate the beginning of the period, in millions of years before the present.

Era	Period	Dates	Canada	Dominant Life
QUATERNARY	Pleistocene	1.65	Surface	Mammoths
CENOZOIC	Pliocene			
	Miocene		Outliers	
	Oligocene		on prairies	Mammals and Birds
	Eocene		on prairies	
	Paleocene	65		
MESOZOIC	Cretaceous	146	Prairies, B.C. Arctic Islands	Dinosaurs Mammals and Birds
	Jurassic	208	Nova Scotia, B.C.	Dinosaurs Mammals and Birds
	Triassic	245	Nova Scotia, B.C.	First dinosaurs
PALEOZOIC	Permian	290	Gaspé, P.E.I.	Early reptiles
	Carboniferous	362	Maritimes, Rockies	Amphibians
	Devonian	408	Manitoba, Rockies	Fish
	Silurian	440		Fish
	Ordovician	510		Fish
	Cambrian	570	Rockies	Invertebrates
PRECAMBRIAN		4600 million	Canadian Shield	Primitive life forms

"TO LOOK FOR THE FOSSILS OF ANY LARGE ANIMALS"

"There were so many bones, you couldn't help walking on them," one collector told me, speaking of his early days in the field in western Canada. Nowadays, even though badland areas are carefully searched by collectors, it is usually easy to find pieces of fossil bone, so in the last century, dinosaur bones must have been very abundant. Since even broken limb bones of large Cretaceous dinosaurs can be several feet long, and many have a shape that is readily recognizable as bone even to the non-specialist, it is reasonable to expect that any traveller in the west who was seriously interested in fossils couldn't have helped finding some. But every collector knows the necessity to "get your eye in" in a new locality — to learn to recognize the colour, shape, and structure of the fossils you are looking for. A number of intelligent, alert travellers — some specifically looking for fossils — came through western Canada in the century before dinosaurs were found, and failed to find any trace of them at all.

Fur trader Alexander Mackenzie (1764–1820) led a group of voyageurs across the continent in 1793. The route led up the Peace River, and when they came to the turbulent waters of the Peace River Canyon (now in northern British Columbia), they found it necessary to portage around it. By detouring they missed one of Canada's most striking dinosaur footprint sites, which has since yielded more than a thousand individual tracks.

A few years later, in 1797, map maker and surveyor David Thompson (1770–1857) was briefed by the North West Company to survey their posts. Among other duties, he was asked "to look for the fossils of any large animals." This intriguing instruction appears to hint that something was already known of the dinosaurs of the west, but the real reason for it is probably quite different. The proprietors of the North West Company were businessmen, and they were no doubt aware of Europe's flourishing trade in mammoth ivory from Siberia. If they could find a source of mammoth ivory in the similar landscape of the western territories, they might have a new source of profit. Although mammoth remains are found in the west and north, no remains have been found in the Canadian west comparable to the complete frozen carcasses in Siberia. During his twenty-eight years in the west (during which he created what has been called "the first comprehensive and accurate map of western Canada"), Thompson searched for bones. Ironically, the closest he seems to have got was in 1809, when his men refused to explore the Athabasca Valley for fear of *living* mammoths! In his memoirs, Thompson described how "all my steady researches, and all my enquiries led to nothing. Over a great extent of these plains, not a vestige could be found, nor in the banks of the many rivers I have examined."

A capable geologist, John Richardson (1787–1865), travelled across western Canada and through the Arctic Islands early in the nineteenth century. He was surgeon and naturalist to two overland exploring expeditions led by John Franklin, in advance of the latter's famous sea voyages to the Arctic. Neither the journeys of 1819–22 nor 1825–27 led through major dinosaur country, though they did traverse parts of the North Saskatchewan River. During the winter of 1825–26, when the party was snowed in at Great Bear Lake, Richardson gave a course in geology to his fellow officers — "almost certainly the first course in geology given in today's western Canada, and perhaps in all British North America." Notes taken by one of his auditors, Lieutenant George Back, have survived, and from these we know that Richardson talked about "animals of the lizard kind" in the Stonesfield Slate of Oxford — Buckland's *Megalosaurus*, the first dinosaur. So half a century before dinosaurs were discovered in western Canada, they were mentioned in a lecture given in what is even now a relatively remote part of the country.

The Palliser Expedition of 1857–60 brought a more formal scientific expedition to the Canadian west, established by the British government to explore communications across the plains and into the Rocky Mountains. Led by Captain John Palliser, its staff included the west's first trained geologist, Dr. James Hector (1834–1907). Although only twenty-three years old at the start of the expedition, Hector was second-in-command. A trained surgeon, Hector had taken some courses in geology at the University of Edinburgh, and had been briefed for his geological work by Roderick Murchison, head of the Geological Survey in Britain. During his years with the expedition, Hector prepared the first description of Canada's geological structure west of the Great Lakes. He was a careful geologist who made maps, measured sections, and kept records. In 1859, the party spent two weeks in the Hand Hills (near present-day Drumheller), during which Hector visited the Red Deer River. He found fossil oysters and ammonites, leaves and wood, yet failed to observe any dinosaur bones! Hector subsequently moved to New Zealand to direct the New Zealand Survey, where he took an active interest in fossil reptiles, even sending bones to Edward Cope in 1890.

The next geologist to miss an opportunity came from the other direction. When founding director Logan retired from the Geological Survey in 1869, at the age of seventy, his successor, Alfred Selwyn (1824–1902), saw an urgent need to reconnoitre the new western territory. In 1873 he travelled overland from Fort Garry to Edmonton, on to Rocky Mountain House, and then returned by York Boat the length of the North Saskatchewan. Although dinosaur bones are not uncommon along parts of that river, Selwyn failed to note any.

FALSE STARTS AND LOST OPPORTUNITIES

The essential foundations of geology were laid in England and Scotland during the first half of the nineteenth century, and because of its practical importance in finding resources for the Industrial Revolution, the science very quickly spread internationally. Dinosaurs were also first discovered in England, but did not have much of a profile anywhere until the vigorous intellectual conflict over evolution brought new attention to the remote past. Although important dinosaur discoveries — particularly of footprints — were made in the eastern United States, almost no one thought of actively looking for dinosaurs until the century was well advanced.

Although Canada quickly adopted the methods of geology, it must be admitted that the discovery of dinosaurs in Canada began badly. *Bathygnathus* and *Arctosaurus,* the early random finds in the east and the north, were thought to be dinosaurs — but weren't — while the real dinosaurs that occurred in those regions were not found until more than a century later. Meanwhile, the abundant remains of dinosaurs on the Canadian prairies evaded keen explorers, some of them actively looking for fossils, during more than a century of exploration. Behind the false starts and lost opportunities, however, somewhat shaky foundations for future palaeontological discovery were being laid.

In the Geological Survey of Canada, the principle of a government-funded, national scientific institution had been accepted, and (despite its focus on basic mapping and economic benefits) the time would come when the Survey would have to take notice of dinosaurs, and also create one of Canada's major museums. While the University of Toronto was hiring cautiously (preferring nepotism and conservatism to inexperience but brilliance), Huxley's successful rival played a part in creating the Royal Ontario Museum. Other institutions recognized home-grown ability, for in 1855 McGill College in Montreal created an opportunity for William Dawson, who became its Principal and Professor of Natural Sciences, and a key player in the Canadian scientific establishment for the rest of his long life.

Of course, as a British colony, Canada was to remain under British scientific influence for some time. However, the United States was a rising star in the scientific world, and with much greater wealth and population than Canada soon built great institutions, staffed by a range of brilliant and colourful scientists. Just as American political and commercial interests knew no boundaries, so its scientists increasingly felt the world was their oyster. Although Canadian dinosaurs were first discovered by Canadians, many of them were collected by American expeditions, and for many years more specimens were to be seen in New York and Chicago than anywhere in Canada.

Within Canada, a more subtle colonialism was apparent. Just as business interests in central Canada collected furs and built railways in the hinterland wherever they wished, so their scientific colleagues saw nothing wrong in carrying off the portable spoils of science to Ottawa and Toronto.

A CANADIAN ADVENTURE

Our stage is set. Within the boundaries of Canada as it approached its modern extent in 1870, a remarkable dinosaur fauna awaited discovery. This book tells the story of those discoveries in the dinosaurs' graveyard, and shows how the past has carried into our lives today.

In the century or so since Confederation, Canada has been shown to have some fifty genera of dinosaurs, making it the fourth-richest country in the world. Many of these genera come from the richest single dinosaur site in the world, Dinosaur Provincial Park in Alberta, which to date has yielded more than 250 skeletons of thirty-six species of dinosaurs, and remains of eighty-four kinds of other vertebrates. Many other sites have been found on the prairies, and more recently in the east, the high Arctic, and as far west as Vancouver Island.

Dinosaur hunters in Canada have included a variety of colourful characters who have made remarkable discoveries. They have been employed by institutions that have slowly developed, flourished, and sometimes declined, fired by imaginative individuals and public demand, but at the mercy of budgets and politics. As the cutting edge of science has returned to dinosaurs in recent years, Canadian specimens, sites, and researchers have played an important part in the new discoveries. Canada's dinosaur science has evolved from an insecure enterprise dependent on outside stimulus, to a thriving cultural industry that not only takes the lead in studying and interpreting the country's own dinosaurs, but also has developed international cooperation with countries as far away as China, Argentina, Morocco, and Uzbekistan.

But why talk just about Canadian dinosaurs? Does it matter where dinosaurs are found? Is there something special or different about them? Certainly some dinosaurs are found only in Canada, and though others are known elsewhere in the world, the best specimens are often from Canadian sites. While Canadian dinosaurs have an important place in international science, the evidence is found in particular places, and is found by people of specific nationalities who have to support themselves and their research. In this sense, science is also a social activity that requires funding, people, and institutions. Just as all history can be seen as local history, so all science is in one sense local science, and deserves consideration as a national (and regional and local) pursuit.

Dinosaurs are not the only important fossils, but unlike most other groups of animals, they have come to be of more than scientific importance. Dinosaurs yield "advantages" Logan never dreamed of, not only scientific, but also educational, cultural, and even commercial. Beyond science, they live another life in our dreams, our books, our art, and our popular culture. They are the foundation of an international tourist and media industry. But in science as in history, it is pertinent to ask: who interprets our stories for us? Can Canadian children find out about the dinosaurs in their own backyards? Do Canadian dinosaurs have a place in the Canadian imagination? Dinosaurs found in Canada are often little known to Canadians, let alone people in other countries. Yet, Canada has built the world's biggest dinosaur museum, and constructed its most spectacular travelling dinosaur exhibit. These stories, too, have a place in this book.

The way in which Canadian dinosaurs have been discovered, studied, and interpreted tells us much about Canadian society, and the way that Canadian resources and achievements are regarded internationally. In the collaboration and tension between British, Americans, and Canadians, easterners and westerners, and scientists and ministers, and in the rise and fall of universities, museums, and programs, we can not only learn something about science, but also about history, and about ourselves as Canadians.

2

"THE MOST IMPORTANT FIELD IN CANADA"

"Sunday June 7th., we waved adieu to what is probably the most important field in Canada, as far as bones of extinct animals is concerned."

THOMAS CHESMER WESTON

THE GRANDFATHER OF THE BUFFALO

Two men gazed at the backbone of a large dinosaur, exposed in tumbled rocks. One was an unnamed member of the First Nations (probably Peigan); the other was Jean L'Heureux, a Québécois, who later wrote down what he saw.

> Among the blocks of erratic boulders are found many fossils of dorsal vertebrae of a powerful animal. These enormous vertebrae measure up to twenty inches in circumference. The natives say that the grandfather of the buffalo is buried here. They honor these remains by offering presents as a means of making the spirit which gave them life, to help them in their hunt.

The time was before 1871, and L'Heureux names the site as One Tree Creek, still used for a stream that flows into the Red Deer River just above what is now Dinosaur Park. L'Heureux's vivid description matches the modern landscape well. There was "a large coulee whose sides to the west are cut almost perpendicularly to a height of more than three hundred feet. . . . One remarkable aspect of the creek is the way it cuts sharply through the bank of the Red Deer River where its bed at the outlet broadens five miles in the form of a semi-circle whose irregular terrain makes one think of the bottom of an ancient lake." In this locality there can be no doubt that such bones must have been from a dinosaur. Though the reported size is somewhat large for an Alberta specimen, his estimate may have been exaggerated by memory.

The Peigan (Pikuni) were one of three nations of the Blackfoot confederacy, whose tepees were often pitched along the Oldman River in what is now southern Alberta. Their predecessors were on the plains when the glacial ice was still melting, and knew the landscape intimately, finding the best places to drive the buffalo, pick berries, and make camp in sheltered places.

They collected internal casts of the chambers of fossil ammonites. Named "iniskim," they were kept in sacred medicine bundles and used in ceremonies to lure the buffalo they resembled. I found records of First Nations' knowledge of mammoth tusks and bones, and could not imagine that they were unaware of the dinosaurs. Eventually I asked Glenbow curator Hugh Dempsey about it, and he referred me to L'Heureux's manuscript.

Born and raised in Quebec, L'Heureux failed to complete his formal education (for reasons rumoured to relate to sexual misconduct), and sought refuge in the less critical west, serving the Peigan as a priest. He helped the Blackfoot in treaty negotiations and prepared a map of the west based on his travels, accompanied by a detailed commentary. Before 1871 he sent these to the governor general of Canada; the notes have survived.

Like other First Nations, the Peigan marked sacred locations by offerings of cloth and tobacco; the "buffalo's grandfather" was distinguished in this way. Although their interest was apparently spiritual, their explanation that the bones belonged to the grandfather of the buffalo is a scientific interpretation by people who knew only one large animal, recognizing both its antiquity and possible relationship to a living descendant. Knowledge only becomes part of the body of science when it is published, so recognition of the First Nations' role has been delayed a century or so. But it is interesting to see dinosaurs incorporated into modern native interpretations of traditional stories (see Chapter 12).

KEY CANADIAN DISCOVERIES TO THE FIRST WORLD WAR

1845 (approx) *Bathygnathus* found, Prince Edward Island

1853 *Arctosaurus* discovered by Osborn, Bathurst Island

1854 Joseph Leidy describes *Bathygnathus*

1871 (approx) Jean L'Heureux reports Buffalo's Grandfather in "Alberta"

1874 George Dawson finds dinosaur bones in "Saskatchewan" and "Alberta"

1875 Andrew Adams describes *Arctosaurus*

1881 Dawson and McConnell find more dinosaur material in "Alberta"

1882 McConnell collects at Scabby Butte in "Alberta"

1883 William Dawson collects in west

1884 Thomas Weston in Cypress Hills and Irvine. Joseph Tyrrell finds *Albertosaurus* on the Red Deer River

1888 Weston sent west to collect dinosaurs

1889 Weston boats down Red Deer River to Dinosaur Park area

1897–98, 1901 Lawrence Lambe collects in "Alberta"

1902 Osborn and Lambe, *On Vertebrata of the Mid-Cretaceous of the Northwest Territory* published

1908 Bensley collects on the Red Deer River for Royal Ontario Museum

THE LITTLE DOCTOR

William Dawson, Nova Scotia geologist and friend of Lyell, continued joint careers in education and palaeontology. In 1856 he was appointed principal of a small Protestant college in Montreal, which soon became McGill University. He was a very modern principal — expert at extracting money from wealthy patrons — but also served as Professor of Geology, giving twenty lectures a week. He worked closely with Logan, for the Geological Survey was based in Montreal, and became an expert in ancient plants. Students referred to social events at the principal's house as "tea and fossils," for they inevitably ended up discussing some new evidence of ancient life that had taken the professor's fancy.

The eldest of his growing family was George Mercer Dawson (1849–1901), who not unnaturally grew up with an interest in the natural sciences. With his sister Anna he collected "plenty of fossils for papa." His parents were devastated when George fell ill at the age of eleven or twelve. For several months he suffered from what his doctors called "a chill" — an illness now known as Pott's disease, or TB spine. After several years of recuperation, his back was deformed, and he ceased to grow in stature, but his mind was as lively as ever. George studied with a tutor, took up photography, and at eighteen spent some time as a part-time student at McGill, though he did not graduate.

At age twenty he started three years at the Royal School of Mines in London, by this time the premier scientific school in Britain. He was taught by Thomas Henry Huxley, dined with the Lyells and other eminent scientists of the day, and received medals and honours each year. His spare time was spent seeing the sights, which perhaps included dinosaur bones at the Natural History Museum. He was offered a teaching post on graduation in 1872, but preferred to return home.

By this time Canada had taken over the Northwest Territories (then including what are now the provinces of Alberta, Saskatchewan, and Manitoba). As Dawson returned, a Boundary Commission was being set up to survey the 1285-kilometre (800-mile) unmarked boundary along the 49th parallel between Canada and the U.S., with teams representing each country. Canada's team consisted of British army officers, astronomers, and engineers, with an armed escort for protection, and there were plans to add a geologist and naturalist to the party. Despite his physical handicaps, George Dawson had the ability, the training, and what in those days was equally important — the right connections. He was appointed to the position; any misgivings his companions might have had were soon dispersed, and he was nicknamed "the Little Doctor."

Life was tough on the Canadian prairies in the 1870s. Travel was by horse or ox cart, nights were spent in a tent or under a Red River cart, and the traveller either carried enough food or hunted for it — buffalo if he was lucky, and grouse or duck if he wasn't. Survival on the road was almost a full-time job, yet a scientist had to fit in his fieldwork during the long hours of daylight, and write up his notes during a siesta break or at night.

Although Dawson was frail and so short that his head barely comes up to the shoulders of his colleagues in expedition photographs, he was tough enough for the job. In later years a miner described Dawson's life in the field: "He was a l'il hunchback runt. . . . When he was in camp he was the first up in the mornin' and off over the hills like a he goat. Be dark when he got back to camp. An' I don't think he never got no decent sleep. Suffered something cruel from thet back of his'n." Others praised his personality: "His constant cheerfulness was a source of surprise to everyone who knew him. . . . In conversation he was witty and humorous to a degree."

Dawson had his own assistant and wagon, but shared a camp with the chief astronomer, Captain Samuel Anderson. The party moved on to the prairies in 1873, and got as far as Wood Mountain in what is now southern Saskatchewan. Large areas of prairie had been burned; others were covered with locusts. They encountered plains grizzlies, and huge camps of Métis buffalo hunters.

In April 1874 Dawson was back in the Wood Mountain area, resuming his work of measuring sections and collecting fossils. In May he found and collected his first dinosaur bones, in a location he describes as thirty-two kilometres (20 miles) south of Wood Mountain settlement, in badlands associated with valleys draining away from the south side of Wood Mountain plateau. In July he found more bones in what is now the Comrey area of Alberta.

Dawson completed a 387-page report on his survey work, which was published in 1875. Dawson needed to refer the vertebrate fossils he had collected to someone qualified to give an expert opinion. Canada had no vertebrate palaeontologist, and London and Owen were a long way off.

In Philadelphia, Leidy had by this time ceased to work on vertebrates, but he had a worthy successor in Edward Drinker Cope (1840–1897). Cope reviewed the specimens, and reported the presence of hadrosaurs, turtles, and a fish. The fossils were deposited with the Geological Survey of Canada, but now cannot be found.

THE GEOLOGICAL CAVALRY

Hot on the heels of the British-led Boundary Survey, Canada's Geological Survey headed west. A transcontinental railway had been promised to bring British Columbia into Confederation, and various routes were being explored. If substantial coal deposits could be found in the west, the railway could run more cheaply. And other mineral resources in the west could help to ensure its viability.

The Survey needed more staff, and one key person was at hand. George Dawson had proved himself on the Boundary Survey, and in 1875 — still in his mid-twenties — he became Chief Geologist of the Geological Survey of Canada. The Survey clearly was buying not only Dawson's skills but also his firsthand knowledge of western geology as a basis for the work that would be done there in the next few years. The Boundary Survey work had given a head start to understanding the plains, so Dawson first spent several years in British Columbia, where the geology was even less well known. In 1881 Dawson was able to return to his beloved prairies, accompanied by a newly appointed assistant, Richard George McConnell (1857–1942), who had recently graduated from McGill University.

Boundary Commission Group.

The western field staff were soon nicknamed the "Geological Cavalry" by the Ottawa head office. Indeed, the west was more suited in some ways to a military operation than a civilian survey. The Mounties had marched west in 1874, but there were few of them, and the still-numerous First Nations were becoming increasingly restless as the buffalo, their staple food, faded away. White whisky traders and petty criminals were present (or at least rumoured to be) everywhere.

By horse, wagon, and canvas canoe the "cavalry" explored the southwestern area of Alberta. Although they carried guns, they found more use for theodolites and geological picks. They were pleasantly surprised to find that the "Indians" were not particularly hostile, although the team was in "constant danger of having horses stolen."

They recorded more dinosaur bones along the Oldman River (then known less attractively as the Belly River). Dawson returned to the new Survey office in Ottawa in the fall of 1881 to write his reports, leaving McConnell to winter in Calgary. In 1883 Dawson (already assistant director of the Survey) assigned McConnell to independent survey, and brought out new assistants, Weston and Tyrrell.

THOMAS WESTON, LAPIDARY

Thomas Chesmer Weston was already in his fifties when he travelled west for the first time. He had learned lapidary — the skills of polishing rocks and minerals — from his father, a mineral dealer in Birmingham, England. Recruited by Logan, Thomas Weston came to Canada in 1858 to begin a 35-year career with the Survey. He had worked in the field in eastern Canada, and had become a capable fossil collector and laboratory technician. In the winter of 1882 he was given the task of "mending and restoring a number of fossil bones collected by Dr. G.M. Dawson and Mr. R.G. McConnell."

Weston brought his own outfit, including his son, G.H. Weston. On June 9 they arrived by train at the end of steel at Maple Creek (now in Saskatchewan). Weston went with Dawson to photograph a Native camp, and was "paralyzed" as a hundred warriors — the real cavalry of the west — galloped towards them across the plain, until it became clear that this was in the nature of a friendly greeting. They headed westward by horse and wagon to Fort MacLeod; Weston enjoyed Dawson's company and looked "back with pleasure to [my] journey with that most courteous gentleman."

Weston then left the main party to work his way back across the plains, and particularly to collect fossils; dinosaurs were surely part of

Scabby Butte, near Lethbridge, Alberta.

the reason for his trip. In the foothills and mountains by Kootenay [Waterton] Lakes, the expedition was impeded by thefts of horses, bad water, dangerous fords, and a broken axle. A lost dunnage bag was returned by "four wild looking Indians mounted on bare-backed ponies" who were rewarded with fifty cents each. Fossil plants and invertebrates were located, and by July 30 Weston's party was finding dinosaur fossils by the Belly (Oldman) River. At Coal Banks (Lethbridge) he found "several bones of a great Dinosaurian reptile, probably the *Laelaps incrassatus* (Cope)."

In the previous year McConnell had found a new dinosaur locality, Scabby Butte, where he collected "a large and interesting collection . . . of reptilian bones, probably of Dinosaurs, some of which are now exhibited in the museum." One proved to be a duck-billed thigh bone, collected in many fragmentary pieces and repaired back in Ottawa by Weston. Scabby Butte is a relatively small area of badlands northwest of Lethbridge. Today it is not obvious to the casual traveller except from a gravel country road that crosses a thin line of badlands, which on my first visit in the 1970s provided access to a "sanitary" landfill — a pile of dead cars and smouldering rubbish. A walk along the outcrops leads into a huge shallow amphitheatre, surrounded by a rim of low badland scarps, cutting into a low hill on the gently rolling prairie.

Weston regarded it as "a remarkable place, reminding one of the excavations of an ancient city. The . . . rocks, have been scooped out . . . leaving portions of stratified rocks standing up like the walls of a burned city." He found teeth of carnivorous dinosaurs, and also collected pieces of fossil wood up to 1.5 metres (five feet) long.

By August 28 Weston's party was back in Medicine Hat, which by then boasted "twenty or more merchants . . . two doctors, two barristers and solicitors, four hotels and several boarding houses." Most importantly, it now had a railway station, and Weston reached Ottawa by September 10. Unfortunately, many of his collections did not reach Ottawa at all, as they were shipped on *Glenfinlas* across Lake Superior — and the ship sank.

DADDY DAWSON'S DINOSAUR

William Dawson obviously had been intrigued by his son's discoveries, and in 1883 he had the chance to see the prairies for himself. In his sixty-third year, he obtained a year's leave of absence from his onerous duties at McGill and travelled to Minneapolis to deliver the presidential address to the American Association for the Advancement of Science.

The railway took him as far as Medicine Hat, on the South Saskatchewan River. Although hardly a geological cavalryman, Dawson explored the neighbourhood, investigating coal deposits, and then travelled towards Ross Creek, which flows north from the Cypress Hills. Here, Dawson was lucky enough to find a "considerable portion of the skeleton of a large dinosaurian reptile." He identified the duckbill remains with *Diclonius*, a name given by Cope to some fragmentary teeth from Montana. Dawson travelled as far west as Calgary, but was not impressed by "a town in a very rudimentary state, its principal hotel being composed of boards and canvas, and floored with sawdust."

It is hard to be certain how much of this dinosaur skeleton went with Dawson, although he did collect an entire buffalo skeleton for the Redpath Museum at McGill. At any rate, the Redpath now only has a fragmentary limb bone from Medicine Hat donated by J.W. Dawson, the only surviving specimen from the first collector in the west from a Canadian university.

IMPERIOUS AND AMBITIOUS

Joseph Burr Tyrrell (1858–1957) was twenty-five when he travelled west, and had a personality that one of his biographers called "heavy-handed, imperious and ambitious." Born near Toronto, Ontario, Tyrrell was the son of a wealthy Irish stonemason. (Various pronunciations of his name have been considered authentic, including TEE-rell — with an accent on the first syllable; or an equal emphasis rhyming with squirrel.) Young Joe learned to shoot but also kept a menagerie of wild animals in his bedroom. Scarlet fever affected his eyes and left him partly deaf; he used an ear trumpet until modern hearing aids were invented after the Second World War, when he poignantly described his emotions at being able to hear the birds sing for the first time in fifty years.

He studied law at the University of Toronto in 1876, while pursuing his scientific interests with courses in meteorology and geology, and (despite his poor eyesight) working for long hours at the microscope. He articled, but soon became bored with the law, and (with the help of his father's political influence with Prime Minister John A. MacDonald) got a temporary position with the Geological Survey. He assisted the Survey's palaeontologist in unpacking thousands of fossil specimens, which provided his only geological training and earned him a permanent position. By 1883 he was out in the west with Dawson, studying the southern foothills and mountain passes.

Tyrrell must have spent many evenings learning from Dawson at the campfire while his chief rolled endless cigarettes and discoursed about the geology of the country they were exploring, but the two did not get on well. Dawson made heavy demands on his own body, and expected his assistants to do the same, leading Tyrrell to complain that Dawson "had every consideration for his horses but no great sentiment about his men."

The party would breakfast and leave by 5.30 a.m. There were no topographical maps, so the party had to create those as well as studying the geology. Tyrrell was given the job of making a pace survey — walking all day while counting his steps, to provide a measured line of march to which geological detail could be related. He was also expected to collect plants for the Survey's herbarium. The work was solitary, as the rest of the party was always behind, ahead, or off to one side exploring a promising exposure of rocks. Food was salt pork, bannock, and tea, and the party took turns to mount guard at night. Tyrrell remembers that he "went out a boy, and came back a man."

As they worked, the prairie was changing. By the end of their first season, the railway had reached Calgary, and the party was able to return to Winnipeg by train, even though they still had to detour through the United States round the Great Lakes to get back to Ottawa.

It was a time of rapid promotion, and by 1884 Tyrrell had an independent command. His mission was to spend three years extending the mapping north from the Belly River to the Red Deer River, over 45,000 square miles (117,000 square kilometres). Although this sounds a formidable task, the area was largely covered by rolling prairie, and there were relatively few geological exposures to inspect. Bedrock was mainly exposed along the rivers, and Tyrrell decided to tackle the Red Deer first. Although he was aware of the dinosaur discoveries further south, there is no evidence that Tyrrell had any particular interest in them, and he was of course unaware of L'Heureux's unpublished report.

A GIGANTIC CARNIVORE

The Red Deer River had been largely bypassed by the fur traders, and was still far from regularly travelled routes. The best access for the traveller was where the Calgary-Edmonton Trail crossed the river, the site of a small settlement that has since grown into the city of Red Deer. From the crossing the river meandered southeasterly, running through a deep valley cut into the plains, to its junction with the South Saskatchewan River near Medicine Hat.

At twenty-six, Tyrrell was in charge of his own party, with three assistants to drive the team, pitch camp, and cook. A canvas canoe helped them to explore the river. Tyrrell followed fur trader Peter Fidler in recording coal (later the foundation of a coal mining industry in the Drumheller area). On June 9 Tyrrell "found and made a small collection of Dinosaurian bones, being the first of such bones found or collected in the valley of the Red Deer River."

On August 11, Tyrrell's party followed Rosebud Creek and Knee Hills Creek down to the valley floor and stayed that night close to the site of the present city of Drumheller, at what was later called Dinosaur Camp, "on a beautiful green sheltered flat with wood, water and grass in abundance to provide for the needs of ourselves and our horses."

> The next day . . . I walked up the bank close to camp, and at an elevation of between forty and eighty feet above the creek found a number of Dinosaurian bones in an excellent state of preservation, though very brittle. . . . among these was a large and fairly perfect head of *Laelaps (Dryptosaurus) incrassatus,* a gigantic carnivore.

In a less formal report, he remembered, "As I stuck my head around a point, there was this skull leering at me, sticking right out of the rock. It gave me a fright."

Tyrrell recognized the importance of his find. "We spent the afternoon excavating these bones from the rock, but unfortunately we had no appliances but axes and small geological hammers." There was some damage, but Tyrrell recovered more than he could carry in his wagon, and had to cache some. "We had," he later reported, "no proper means of packing them, and no boxes but the wagon box to put them in. . . . After completing this work, and packing up our precious collection as well as we could, we started in a cold drizzling rain. . . . Our journey to Calgary took us a week, for we were obliged to drive slowly and carefully . . . over the rough unbroken prairie. . . ."

Tyrrell never forgot that early experience, and it fascinated him enough that he wrote an article about it forty years later, in 1923. Yet no one would have been more surprised than Tyrrell to learn that this find is, for many people, the main reason he is remembered today. For Tyrrell was not a palaeontologist, and did not make any other dinosaur discoveries — and would be embarrassed to find himself described inaccurately as the man who found the first dinosaur bones in Canada, or "discovered the rich dinosaur beds of Southern Alberta."

Tyrrell's long and distinguished career include many other achievements, as an explorer, a mining engineer, and a historian. His dinosaur reputation comes from an historical accident: a century later the Alberta government built a new palaeontology museum close to the site of his find, and chose to name the museum — now the Royal Tyrrell Museum of Palaeontology — after him. It is understandable that its millions of visitors now associate the name Tyrrell first with dinosaurs, and are unaware of several people with far greater claims to such recognition.

"COLLECTING DINOSAURIAN BONES"

The Geological Survey's only fossil collector was kept busy in eastern Canada, so Tyrrell's discovery was not followed up on for four years. However, Weston did prepare collections from the west during the winters, and in 1888 he travelled west again. By now in his mid-fifties, he remained a nervous traveller, cowed by coyotes and rattled by rattlesnakes. However, he was a steady worker who got to where he was going — if not always when first planned. As a result of Tyrrell's "suggestion and solicitation," part of his expedition was "for the especial purpose of collecting Dinosaurian bones" in the vicinity of Tyrrell's earlier finds on the Red Deer River.

After a visit to the Cypress Hills he headed for Calgary, and took the stagecoach north, arriving at Red Deer Crossing on August 5. During the winter Weston had arranged with the Rev. Leo Gaetz for a boat to be built, but it was barely started. On August 13 the craft was launched, and after dinner with Rev. Gaetz, Weston and two new helpers, with supplies for three months, "embarked in our rudely constructed boat, and drifted down the swift current." However, Weston found that his "men seemed to know nothing about navigating rapid streams," and abandoned the voyage when a Métis farmer, McKenzie, offered "to have another boat built by the following spring and to accompany me himself." On the way home Weston regretted the loss of his adventure, "but as McKenzie said we could not afford to lose our lives for a little geological work."

In the summer of 1889 Weston made the first dinosaur-hunting voyage down the Red Deer River. On the train Weston had the company of George Dawson, now director of the Survey, who was on his way to British Columbia. Weston's new companion, McKenzie, "a fine example of the half-breed Indians of the Canadian North-West Territory," met Weston at Calgary, and they arrived at the farm on June 13.

Weston left with McKenzie and his son in two boats on June 17. The river was "a smooth stream of clear blue water two hundred feet wide," which soon passed through a narrow canyon full of boulders. Soon they were in calm water, and Weston had leisure to enjoy the majestic scenery.

> Bold escarpments several hundred feet high occupy portions of the river-sides for many miles, and from the table-land, sloping valleys, thickly wooded in places, form lovely retreats for the numerous wild animals that roam the sheltered recesses.

Weston noticed coal seams; red and yellow burnt shale where coal seams have been set alight through lightning, grass fires, or spontaneous combustion; tracks of grizzly bears; and abundant beavers. By June 23, they were "fairly in the bad-land district of the Red Deer River. The great sandy buttes and escarpments . . . extend for many miles, sometimes coming close to the river side, and in other places lying back, leaving room for grassy and sandy flats." Here Weston began finding dinosaurs. He located the skull and other bones of a carnivore, and in one jaw the teeth were "almost perfect — large, curved and beautifully serrated at the edges. There were also portions of limb-bones and claws — dreadful claws — the sight of which carried one back into past ages." Slabs of sandstone preserved such ephemeral traces as ripple marks and rain-prints, and there was a variety of plant remains.

Weston passed through the area now occupied by Drumheller, where Tyrrell had found the first dinosaur of what became known as the Edmonton Formation. He sailed on into "verdant valleys" representing marine beds of what became known as the Bearpaw Formation, and on into new dinosaur-bearing layers where "sandy buttes and escarpments are strewn with fossil bones of dinosaurs and turtles, and scales of fish."

The boating was not all idyllic. "In some parts of the river . . . our boat . . . frequently stuck fast in the sands," and it was not easy to get to deeper water "as our boat pole sunk its full length without finding bottom." The land also had its perils, for when Weston crossed an apparently dry creek bed he sank "up to my waistcoat pockets" and pulled himself out with difficulty. McKenzie failed to touch bottom at that spot with the 4.5-metre (fifteen-foot) boat pole, and commented dryly, "We should have knowed where you had went to, for your hat would have been left."

Relief from peril was celebrated on Dominion Day, when the party raised a red pocket handkerchief on a pole, fired a salute, sang "God Save the Queen," and drank to all absent friends and relatives "by tapping our only bottle of brandy, which I had guarded diligently in case of getting a rattlesnake bite."

"THOUSANDS OF TONS OF FOSSIL BONES"

Soon Weston's party entered Dead Lodge Canyon, and he recorded the first fossil collector's impression of what is now Dinosaur Provincial Park.

> The river cuts through a fertile valley from 400 to 600 feet deep. Here nature has used her scooping shovel to an enormous extent, for between the prairie level and the river sandy buttes interstratified with bands of sandstone form pyramid-like structures.

There is a hint of the richness to come, for "here to my delight I found on the bleached sands numerous fossil bones." A carnivore leg bone was found, but Weston's techniques were not adequate. After three hours were spent digging it out, "three pairs of hands carefully lifted our precious specimen . . . when to our surprise the thing crumbled into a thousand fragments."

Weston recognized the richness of the site: "In these great sand banks must be buried thousands of tons of fossil bones, for as the weathering away goes on, these relics . . . weather out and remain on the sandy flats or roll to the foot of the banks and cliffs." He was also aware that the rocks were of the same age as those he had already explored further south, and thus older than the rocks in which Tyrrell's find had been made; this was the first recognition of the beds that have been called over the years the Belly River, Oldman, Judith River, and now Dinosaur Park Formation.

Food was running out, and with great reluctance, Weston boated down as far as the "Battleford and Swift Current Crossing," where there was a shack occupied by the Mounties. Here they secured a freighter's cart, and shifted a "dozen or more boxes of fossils weighing about a thousand pounds."

In his later assessment Weston clearly identified the future Dinosaur Park area as "the most important field in Canada, so far as bones of extinct animals is concerned." Weston never saw the Red Deer River again, but he had pioneered the most effective method of travel, distinguished two formations, collected a significant amount of dinosaur material, and identified the priority for future work. "It would take years to glean from these great sandy buttes, flats and cliffs even a part of the information they hold relating to that period when the Cretaceous rocks of Canada were laid down," he concluded. More than a century later, the richness of the site is still being explored.

George Sternberg plasters Chasmosaurus.

FLOUR, PLASTER, AND FIBREGLASS

"The most important geological specimen found . . . was a bone three feet long, eight or ten inches wide and half an inch . . . thick. I spent six or eight hours uncovering this bone, and Mr. Macoun walked sixteen miles — to Medicine Hat and back — to get glue or some other material with a view to preserving this specimen. It was like most of these fossil bones, cracked in all directions. Well after all our trouble, while lifting it from its sandy bed, it fell into a thousand fragments, and now lies at the bottom of one of the great excavations in these soft sandstones. Still I mourn the loss of this bone."

THOMAS CHESMER WESTON

Weston's problem was all too common in the early history of dinosaur discovery. Although some bones are heavily petrified, and can be handled easily, such fragile bones are abundant in the Canadian west. Palaeontologists use a standard technique, which is essentially that used by doctors to protect a patient's arm or leg while a broken bone repairs itself.

As the bone is excavated from the rock, it is soaked with shellac or a glue in solution to stick loose bits together and strengthen the bone itself. Then (usually over a coating of tissue paper) the upper surface is coated in burlap bandage dipped in liquid plaster of Paris. Once this has set, the bone is undercut, and rolled over. More rock is removed to reduce the weight, then plaster can be applied to the underside, enclosing the entire bone in a plaster jacket. With very large bones, wooden splints strengthen the whole package.

This technique is convenient, and uses materials available from local hardware stores. It allows plastered bone to be carried safely over rough ground, and protects it when bounced in a truck or railway car to its destination. Although modern materials such as fibreglass are now occasionally used, the principle is the same, and the technique has been part of vertebrate palaeontology for around a century.

The origin of the method is not clearly established. Plaster was recommended for strengthening fossils in England as far back as 1836, and it continued to be used there for a while. Bandages dipped in plaster of Paris were first used for orthopedic purposes by a Dutch army officer in the early 1850s, and the technique was in use by a New York surgeon a few years later. The method seems to have been independently applied to fossils by several people in the American west in the 1870s. Cope and Sternberg used rice paste with burlap in 1876, and Samuel Williston pasted stiff paper onto bones in 1877. In the same year, Arthur Lakes seems to have been the first to use plaster for dinosaur bones, and the modern technique is considered to have been in general use in the 1880s by Marsh's collectors. However, some of the American Museum collectors were still developing it in the 1890s.

Weston and Macoun were clearly trying to develop suitable techniques, for Macoun was looking for glue. Yet the inventiveness of capable technicians in the Canadian west did not come up with the plastering technique. It is particularly puzzling that Lambe did not learn the technique when he was in New York. In the early days of dinosaur collecting, sites and methods were kept very secret, so perhaps this was the problem. But surely Lambe would have seen some plaster jackets in the stores or laboratories.

Without the method, only hard, solid bones could be successfully collected. And the technique did not come to Canada until Barnum Brown arrived on the Red Deer River in 1909.

NO OSTEOLOGIST

"There being no osteologist connected with the Canadian Geological Survey, it was decided that this . . . collection should be placed in the hands of Professor E.D. Cope, of Philadelphia . . ." comments Weston during the winter of 1884–5. He was discussing his, Dawson's, and McConnell's mammal collections from the Cypress Hills, and an osteologist — someone who studies bones — was Weston's way of distinguishing what we would now call a vertebrate palaeontologist from someone who studies fossil shells or trilobites, corals, or plants.

For a decade, Dawson, McConnell, Tyrrell, and Weston had all been collecting fossil vertebrates, and some of the fragile bones made it back to head office. The Survey had a number of capable palaeontologists on staff, but could not cover the entire field. Fossils were made available to specialists in Canadian universities for study and description; for instance, William Dawson frequently reported on the fossil plants. The Survey had its material identified, while the specialists could continue their research without expensive and time-consuming fieldwork. Though some fish, amphibians, and even primitive reptiles had been described by William Dawson, J.F. Whiteaves of the Survey, and a New Brunswick part-timer, G.F. Matthew, Canadians were without experience of dinosaurs, and the country had no collection of bones that could be used for comparison.

At first, the solution was to look outside the country for expertise, as had happened with *Bathygnathus* in the 1850s and Dawson's collections from the 1870s onwards. Before the end of the century, vertebrate fossils from Canada were described or identified by Ray Lankester, Richard Owen, and Arthur Smith Woodward in London, and Alexander Agassiz, Edward Cope, Joseph Leidy, Othniel Marsh, and others in the United States.

Further finds by Dawson, McConnell, and Weston were referred to in a report by Dawson and McConnell published in 1884. There were many references to dinosaur bones, but nothing is identified except "a detached tooth," which Cope thought may belong to his carnivorous genus *Laelaps.*

By the winter of 1884–5, extensive collections of dinosaur material were also referred to Cope. "The Professor was invited to come to Ottawa and select such specimens as he thought worthy of description and illustration. Prof. Cope came, made his selection, and they were shipped to his rooms in Philadelphia." By this time, the young man who would became Canada's first dinosaur palaeontologist had joined the Survey.

A LESS-DEMANDING CAREER

Lawrence Morris Lambe (1863–1919) was the son of a Montreal advocate and his wife, Margaret Morris, of English and Scottish descent. He graduated from the Royal Military College in Kingston, and while waiting for a military appointment he worked as an engineer building the railway line through the Rockies. There he caught typhoid fever, and looking for "a less demanding career," joined the Geological Survey as assistant palaeontologist in 1885. Initially working as an artist, he threw himself into research, at first on corals. "Palaeontological work," his colleagues remembered, "was to him indeed a labour of love. The little worries of life seemed never to penetrate his optimistic temperament. His friends will long remember the cheery smile and kindly word with which he always greeted them."

After Weston's expeditions, the Survey ignored the Red Deer River for nearly a decade. Director Selwyn retired in 1895, to be replaced by George Dawson, whose more personal interest in the dinosaurs perhaps led him to send Lambe to the the Red Deer River in 1897. The trip west allowed Lambe to observe a test boring for oil and gas, but he then boated down the river for a month, paying particular attention to the Berry Creek and Dead Lodge Canyon areas, "the main object being to make a thorough search for dinosaurian and other organic remains in the rocks of the Belly River Formation." His finds included a variety of isolated vertebrae and broken limb bones, including "a very large tibia, four feet long."

The next year, 1898, Lambe left Ottawa in July, and with two local assistants travelled by wagon from Medicine Hat. His heavy wagon, laden with a 3.5-metre (12-foot) spruce boat, struggled over muddy trails. They camped opposite the mouth of Berry Creek, near the site of what later became the town of Steveville, on the western edge of what is now Dinosaur Provincial Park. Lambe worked the area for a month, spending half his time on each side of the river.

The broken terrain made collecting very difficult, and the wagon could not be used for more than a mile or so from camp. Bones "had to be carried to camp in an improvised stretcher often for some miles." His specimens were largely single bones, including "well preserved separate parts of jaws, horn cores, bony scutes . . . vertebrae, ribs, a perfect sacrum . . . limb-bones, the largest of which is a femur four feet in length . . ."

The closest he came to a complete skeleton was found "within an area twelve feet square, a number of bones representing the remains, no doubt, of a single individual . . ." These included "an almost entire fore-leg" whose bones were "found together in their proper relative order and indicating an individual of large size." There were also ribs, vertebrae, and jaw fragments, and "some of the large bones of the hind legs, but it was impossible to remove them as they were in a crumbling condition."

REFERABLE TO *TRACHODON*

In the spring of 1897, Cope died suddenly, on a cot in his office, surrounded by the fossils that had dominated his life. As Lambe toiled in the field, he must have wondered who else could identify his material? Othniel Marsh, Cope's rival, was still alive, but he was in his sixties, and his reputation had been damaged by controversy.

Cope's obvious successor was Henry Fairfield Osborn (1857–1935), who had been at the American Museum of Natural History in New York since 1891. But Osborn had not at this time published on dinosaurs, nor had the museum acquired Cope's collection. The Survey decided to develop Lambe's interests in vertebrates, and Osborn was requested to "exercise a general supervision over this work as it proceeds." Although he provided some training, Osborn insisted in a letter to Dawson that since "the lion's share of the actual work" must fall to Lambe, "he should receive the fullest credit."

Lambe's work on corals had shown him how to identify specimens and describe new species. However, dinosaurs were not as easy to study. "The difficulty of arriving at a proper understanding of the generic and specific relationships of many of the bones to each other, and of their affinities, is enhanced by the very scattered state in which they are found," he commented. Even if a species was already known to science, descriptions were poor, illustrations were inadequate, and individual bones were much alike. Lambe at first tried valiantly to relate his finds to the nearest kind of dinosaur already described from the U.S. or Britain, running into all kinds of pitfalls, for we know now that most of the dinosaurs he was finding were unknown to science.

In his 1898 report he suggested that "the majority of the larger bones . . . are referable to *Trachodon.*" However, he found a horn core with a jaw fragment "with teeth of the *Trachodon* type" and boldly concluded that "the species of *Trachodon* here represented . . . had well developed and formidable horns." *Trachodon* was in fact a poorly defined name given to mixed teeth of duckbills and horned dinosaurs found in Montana. In later years it was interpreted as a duckbill, but Lambe is not necessarily suggesting a duckbill with horns — he had probably recognized the similarity of his find to the horned dinosaur teeth in the original description.

He also identified what he considered remains of the horned dinosaur *Triceratops,* as well as teeth of the carnivore *Laelaps,* and part of the skull of the armoured dinosaur *Nodosaurus* — representing the most conspicuous groups of Alberta dinosaurs. Much more collecting and careful description was necessary before it was recognized that many new dinosaurs were being found.

A SPECIAL DEPARTMENT OF RESEARCH

The American Museum of Natural History, on its site beside Central Park, was already a landmark in New York City. Although the museum had been founded in 1869, Osborn did not start its Paleontology Department until 1897, while also teaching at Columbia University. Osborn had travelled to the western United States to collect fossils in 1877–78, studied in London under Huxley in 1879, and published important work on fossil mammals. He was square and moustached, self-confident and self-important, brilliant and communicative. He was also wealthy — he mentions casually in a discussion of great men that he had the privilege of calling the millionaire J. Pierpont Morgan "Uncle Pierpont."

Osborn used his connections to secure funding, purchased Cope's dinosaur collection from his widow, and attracted some of the finest palaeontologists and fossil collectors in the world, beginning a program of dinosaur collection and research that soon made the museum a world leader in the field. In June 1900 Lambe spent a week in New York to study with Osborn, "who afforded him every facility to study the collections and to familiarize himself with the methods employed there."

In February 1901, George Dawson died quite unexpectedly. He was only in his fifty-second year, and had been absent from work for only one day. He was succeeded by Robert Bell, an expert geologist already in his sixties, who (to his frustration) remained as acting director for five years. Bell increased the Survey's budget substantially, promoted fieldwork in new areas, and turned his attention to publishing maps and reports for which the fieldwork had been completed. Dinosaurs were no longer a priority. Lambe had only one more field season on the Red Deer River, in 1901. He spent three months using the same base but exploring downstream as far as Dead Lodge Canyon. His collection was the best yet made from the western Canadian Cretaceous.

Lambe, who was only in his late thirties, worked on alone in the laboratory without assistance. He was now custodian of the largest collection of dinosaur and other vertebrate fossils in Canada, many of which were potentially new to science. "Lambe was a good scientist, a careful observer, and a diligent recorder of facts," and he began, slowly, step by step, to disentangle what we now know to be faunas almost totally new to science.

During a second visit to New York in February to April of 1903, Lambe took Osborn's "special post-graduate courses" in vertebrate palaeontology at Columbia, and studied "the magnificent collection of vertebrate remains" in the American Museum of Natural History. He also examined other collections at the U.S. National Museum in Washington, D.C., the Carnegie Institute in Pittsburgh, and the university museums at Yale and Princeton. Osborn, in a letter to the minister in charge, strongly recommended Lambe's formal appointment as a vertebrate palaeontologist. "Great discoveries have been made in the Northwest Territory, and will be made in the future," Osborn urged. "[I]t appears of the utmost importance for the . . . great museum which has been approved by your government, that vertebrate paleontology should form a special department of research and be cared for by an able specialist." Lambe may have solicited the letter of support to increase his income, for when he was appointed he married Miss Mabel Schreibe.

Lambe's early reports were soon supplemented by scientific papers in Geological Survey memoirs and scientific journals. In 1902 he published the first detailed account of his findings, in a report entitled *On Vertebrata of the Mid-Cretaceous of the Northwest Territory.* An introduction by Osborn distinguished two dinosaur faunas, the highest (most recent) in the Edmonton Formation near Drumheller, and the lower (oldest) in the Oldman Formation (Belly River series) near Steveville. Lambe's material was still very limited; for instance, he did not have any complete skulls to work with, and only a couple of partial skeletons had been recognized in an area that we know produced many complete ones in subsequent years. Nevertheless, he described a number of dinosaurs (for instance naming three species of the horned dinosaur *Monoclonius).*

Between 1899 and his death in 1919, Lambe published one or more papers on vertebrates each year, describing new genera and species, and elaborating on new features of those already named. As a later researcher, Peter Dodson, points out, "Lawrence Lambe was the first Canadian scientist to collect [dinosaur] fossils in Alberta, to recognize their significance, and then to describe them."

PROGRESS OF VERTEBRATE PALAEONTOLOGY

By 1900, in the quarter century since G.M. Dawson first documented dinosaur bones in western Canada, dinosaurs had been recognized from several localities. The Red Deer River dinosaurs had been shown to belong to two different formations, and the importance of Steveville had been recognized. Techniques had been devised to penetrate the remote area in which the fossils occurred, even if material could not be collected undamaged. Many specimens had been collected by the

Geological Survey, and a token investigation made by McGill University. Although Dawson's initial discoveries were made on a British-sponsored expedition, all later work had been done by Canadian expeditions, initiated by Canadian institutions, and carried out by Canadian-born (and largely Canadian-trained) geologists. The only American involvement to date was expert assistance in identification by Cope and assistance by Osborn.

However, dinosaurs did not occupy a central place in the Survey's concerns, and no other organization had emerged to take over this role. At first Lambe tried to justify work on dinosaurs by the potential economic value of the information, commenting that "The importance of a more intimate knowledge of the fauna of the Edmonton Series is apparent when it is borne in mind that the beds of this series in Alberta constitute the principal coal-bearing horizon of the district." However, in reporting to the Royal Society of Canada in 1904 he felt no need to apologize for studying dinosaurs, indicating that "special attention has been given . . . to . . . rocks as exposed at [the] . . . Red Deer River . . . in an endeavour to obtain collections as representative as possible of their fossil vertebrate remains."

Lambe lists *Bathygnathus* as a Triassic dinosaur, and *Arctosaurus* as a carnivorous dinosaur from Mesozoic beds of uncertain age. All other Mesozoic reptiles listed are from western Canada. The lower beds (Belly River series) are much richer, having "a greater diversity of forms and a larger number of species." Here Lambe lists an impressive twenty genera (nearly thirty species) of reptiles.

Of the Belly River fauna, Lambe regarded eight genera as dinosaurs. They included both the heavy carnivores *Deinodon,* and the lightly built bird-mimic *Ornithomimus.* Lambe also lists the little *Troodon* (now known to be a small carnivorous dinosaur) as a lizard. Two genera of armoured dinosaurs (*Palaeoscincus* and *Stereocephalus*) are listed. Horned dinosaurs included Lambe's species of *Monoclonius. Stegoceras* (which we now know as the first of Canada's bone-headed dinosaurs) was known only from "portions of the skull," and Lambe interpreted the great thickening of the forehead as the base of a horn, producing "an entirely new type, a unicorn dinosaur remarkable in that it bore a horn springing from the fronto-nasal region." Five duckbill species are listed, four as species of *Trachodon*, and one other genus, *Cionodon* (poorly based on teeth described by Cope from Wyoming). As well as dinosaurs, Lambe includes a number of turtles, a plesiosaur, two crocodiles, and the crocodile-like champsosaurs. As well as reptiles, some fish, an amphibian, and a couple of mammals are recognized

— the beginning of understanding the rich fauna among which the dinosaurs lived. From the youngest beds (Edmonton series) Lambe lists only the carnivore *Dryptosaurus.*

The different kinds of dinosaur that had been recognized belonged to four of the six major groups recognized by the end of the nineteenth century. Saurischian dinosaurs recognized in Canada were all carnivores; there were none of the giant sauropods that were astonishing collectors and the public in the western United States. Most were ornithischians, of which the most common were duck-billed dinosaurs (hadrosaurs), and some were horned dinosaurs (ceratopsians). The armoured dinosaurs were at that time lumped in with the plated dinosaurs such as *Stegosaurus,* known from the American Jurassic.

3

THE CANADIAN DINOSAUR RUSH

". . . nothing can be done but beat them to the best specimens."

W. D. MATTHEW

A MINIATURE GRAND CANYON

In the spring of 1909, an Alberta rancher, John L. Wagner, went to see one of the sights of New York City, the American Museum of Natural History. He was interested in the dinosaurs on display, because (as he pointed out to the staff) there were bones just like theirs on his ranch, on Michichi Creek near present-day Drumheller. Through Lambe's work with Osborn the museum staff were of course already aware that dinosaurs were to be found on the Red Deer River. One collector, Barnum Brown, was in Montana the following summer, and made an exploratory visit to Wagner's ranch before returning to base. What he saw began the period often called the Canadian Dinosaur Rush, on the analogy of the gold rushes that attracted so many men to remote places.

In 1910 Brown, newly appointed Associate Curator of Vertebrate Paleontology, headed into Alberta for his first full season, with colleague Peter Kaisen and other collectors. The party travelled by train to Calgary, where Brown took pains to pay respects to William Pearce of the Canadian Pacific Railway. Pearce contacted the railway's freight agent, explaining that Brown would be shipping fossil bones, and adding (to clear up any possible misunderstandings), "This is not perishable freight."

Brown improved on Weston's boat technique by employing several carpenters to construct a 3.5m x 9m (12′ x 30′) raft to float down the Red Deer River — big enough to carry a load of ten tonnes. On its deck stood a large wall tent containing a sheet iron stove, vented through a black chimney. The raft could carry a season's food, plaster, and lumber — and large fossils.

The first hundred kilometres (sixty miles) below Red Deer were known as The Canyon, where the river ran faster than the usual six kilometres per hour, and Weston's 1888 trip had been abandoned. After an exciting passage, Brown's party safely passed Erickson's Landing, where an enormous landslide narrowed the stream. Beyond, the scow drifted into a deep valley, which Brown described as "a miniature grand canyon." He later described the landscape, with the spruce and poplar trees, and ripe red raspberries colouring the underbrush. Few buildings could be seen on the plateau above, and the silence was only broken by flocks of ducks and geese, and the hum of the legendary Alberta mosquitoes, which forced the collectors to wear gloves and protect their faces with nets. Through the long days they drifted down, tying up at likely places to look for bones. At night horned owls hooted, and coyotes yipped from the bluffs. After a strenuous day's work, the collectors enjoyed supper, followed by a pleasant hour before sunset fishing or exploring in the boat. Later, pipes were lit round the campfire, and tales of other collecting trips were shared.

BARNUM BROWN

When Barnum Brown joined the American Museum of Natural History in 1897, there was not a single dinosaur in its collection. At his death in 1963, it had the largest and most important dinosaur collection in the world. When he came to Alberta, Brown was already an important dinosaur collector, with more than a decade of experience.

Barnum Brown was born in Carbondale, Kansas, in 1873, a few days before the great showman Phineas T. Barnum's "Great Traveling World's Fair" arrived in town. Brown was named after the great showman, and also lived his life with a certain panache. His father, William Brown, both farmed and ran a coal mine. Young Barnum collected fossil shells turned up by the plough and learned about rocks in the mine. He started serious fossil collecting as a student at the University of Kansas, and in 1895 found his first dinosaur, a skull of the three-horned *Triceratops.*

He joined the museum staff in 1897, and was so busy in the field he did not finish his degree until 1907. Brown was sent to excavate a *Diplodocus* skeleton in Wyoming, then (at an hour's notice) to Patagonia, eventually returning (via Europe) only in 1901 (missing Lawrence Lambe's first visit to the museum). Osborn thought very highly of him, and said:

> Brown is the most amazing collector I have ever known. He must be able to smell fossils. If he runs a test-trench through an exposure it will be right in the middle of the richest deposit.

Brown was an elegant dresser even in the field, a fine dancer who was always something of a ladies' man, and a careful manager of press and public relations. One first impression was remembered: "Dr. Brown was tall, straight, deliberate, and thorough, with twinkling blue eyes that went well with pince nez. His features suggested a scholar rather than the field explorer, being a bit on the dignified side." He was nicknamed "Mr. Bones" and became something of a celebrity in the field, with ranchers and townspeople welcoming his "travelling circus," and falling over themselves to provide camping places and other facilities.

In 1902 he took his first of several expeditions to Upper Cretaceous beds in Hell Creek, Montana, where he found what became the most famous carnivorous dinosaur, *Tyrannosaurus rex.* In 1904 he married, but his wife died in 1910. The heartbroken Barnum accepted an offer from his parents-in-law to raise his child, and headed for the field to forget his sorrow in hard work.

His colleague Roy Chapman Andrews described how Brown would disappear from the museum for parts unknown, ". . . invariably his whereabouts was disclosed by a veritable avalanche of fossils descending in car-loads on the museum. Barnum himself would follow eventually, just drifting in quietly with no fanfare."

Thomas Chesmer Weston.

"SOME NEW CREATURE OF BIZARRE FORM"

Near Big Valley (about eighty kilometres — fifty miles — east of Red Deer) dinosaur bones became abundant. Fossil localities were named in relation to tributary streams and ferries, almost the only landmarks along the river. Brown made the first use of modern collecting techniques in Canada. He looked for bone fragments at the bottom of the buttes, and followed trails of fragments up the slopes until bones were found in the rock. Then "we follow in from the exposed surface, uncovering the bone with crooked awl and whisk broom, careful not to disturb the bone itself."

Once a decision was made to excavate, Brown moved in the heavy equipment. With "pick and shovel the heavy ledges above are removed, and often a team and scraper and dynamite are used when a large excavation is to be made." Closer to the bone, more care was taken. The plastering technique was used for the first time in Alberta, and the blocks were packed in hay, in specially constructed boxes. Three days would be enough to uncover a skeleton, but it could take three weeks before it could be boxed.

Five kilometres (three miles) above Tolman Ferry the party found what would became the type of a small horned dinosaur *Leptoceratops gracilis,* in what is now the Scollard Formation. The bones weathered out on an old cow trail, many too fractured for repair. Later Brown wrote: "Any prospect may reveal some new creature of bizarre form, and we are constantly finding skeletons of animals known before by parts only."

Brown had kept in touch with Osborn, who wrote back, "I congratulate you and the Museum with all my heart. Now you must keep quiet about this find, following Marsh's adage not to go hunting with a brass band, and get what we need of this wonderful fauna before other explorers find their way there." In September, ice formed on the river, and it became too cold to collect. Brown reached Tolman Ferry, unloaded his "ark" and beached the boat, so he could continue his journey downstream next year.

A PERFECT BEAUTY

In 1911 a few bone fragments on a hillside near Tolman Ferry led Brown's party to a complete skeleton. At first thought to be a *Trachodon* (a crestless duckbill), its skull proved to have a long crest. It was preserved in ripple-marked sandstone with worm tracks and horsetail impressions, and with the skeleton and fragmented skull were bits of skin, suggesting quiet burial on a sandy beach. The 320-kg (700-pound) box containing the fossil had to be lowered by block and fall to the valley below. The species was later named *Saurolophus* ("crested saurian") *osborni,* honouring Professor Osborn, who saw this site on a visit.

Osborn perhaps also saw an ankylosaur quarry, high up on the face of a steep cliff, where a 9 x 12 x 6-m (30 x 40 x 20-foot) hole had to be blasted. Brown had found this the year before, and contracted out some of the work to be done during the winter to a local rancher. However, Brown's crew had to finish the job in 1911, eventually moving nearly nine hundred cubic metres of sandstone — and finding the specimen still incomplete. Even so, it was important. "The *Ankylosaurus* skull is a perfect beauty, uncrushed, all plates in position," Brown wrote enthusiastically.

Brown took Osborn on a 240-km (150-mile) canoe trip onwards down the river. This allowed Osborn to view the rest of the terrain exposed by the Red Deer River, and showed him the region he had written about with Lambe. Brown and Osborn continued down to Drumheller, sampling the beds at intervals, then passed through the marine Bearpaw Formation. Then they saw for the first time the wonderful badlands which are now partly protected in Dinosaur Provincial Park. Here, they knew, were rocks older than those on which Brown had been working, and of equal importance for dinosaur discovery. At the rate he was progressing, it might take several more years to work the entire river.

In 1912, Brown was back again in the Edmonton Formation. He added a powerboat to his navy, but its technology was primitive. "The motor boat is a success and I am gaining experience as an engineer," he wrote to Osborn. "It shows more varied moods than a woman, but I still maintain a Christian spirit towards it." He collected "eight skeletons from a limited area exposed along three miles of the Red Deer River," and could continue his leisurely pace indefinitely.

ASK THE SUPREME COURT?

"The people of Canada began to say 'What's the idea, we're letting all our dinosaurs go to the United States,'" remembered Charlie Sternberg about this period. As Brown's activities became better known, public pressure mounted on the federal government to take action. The politicians in turn passed the buck to the Geological Survey for action — but what action?

Colonial Canada had been surveyed by British expeditions, and as long ago as 1882, William Dawson commented on "the magnificent and costly surveys and commissions of the United States, which freely invade Canadian territory whenever they find any profitable ground that we are not occupying." In this way British and U.S. agencies had done useful scientific work in Canada, publishing results that were available to underfunded Canadian agencies, even if the specimens had been removed. This tradition of scientific colonialism was being continued by Brown, and though he cultivated influential people in the "host" country, he was not able to stop complaints to the government.

As the Survey no doubt pointed out to their political masters, Lambe had compiled a good list of fossil vertebrates from the region — yet Brown was clearly finding new dinosaurs. Survey managers may have felt that dinosaurs were of no practical importance, yet — now that Canadians for the first time were beginning to take a serious interest in their own dinosaurs — it would be a pity if they were all removed before any further work could be done. Moreover, the Survey was now planning a new National Museum, which could use some dinosaurs for display. No one knew how many dinosaur fossils were available, or for how long Brown intended to collect.

Some agonizing questions must have been discussed in Ottawa offices. Would Brown leave any fossils? Should his large-scale exporting be stopped? Since Canada was not doing any collecting itself, could there be reasonable objection — especially when the Survey was busy acquiring fossils from the western United States for its new museum? If objection was to be made, what administrative or legislative route was available? (Charlie Sternberg thought that the Supreme Court was asked, and stated that such a law would be unconstitutional — surely unlikely since such laws are now widespread.) What would be the international repercussions if Brown was forbidden to work in Canada — would it be regarded by the U.S. as an unfriendly act? And if Brown was prohibited, could Canadians then ignore the fossils themselves?

PROTECTING FOSSIL SITES

Compared to historic buildings and the more spectacular archaeological sites, governments have generally been slow to protect fossils. In the United States the first site to be protected was Dinosaur National Monument in 1915. However, precedents in other fields could have been used. By 1900, legislation had been established to protect archaeological remains in a number of countries, though it was not always well enforced. By the time Brown headed for Alberta, American archaeologists had encountered problems in Peru, where excavation was forbidden without a permit as early as 1893. Legislation of this kind could have been enacted for fossils.

The U.S. itself took a different approach when it established Yellowstone National Park in 1872 — the first of what is now a worldwide network of national parks. In Canada, Brown's contact William Pearce had urged protection of Banff in 1885, eventually leading to Canada's first national park in 1887. By the time Brown arrived on the Red Deer River, western Canada had several national parks. Although in theory the badlands of the Red Deer area could have been made a national park, there is no evidence that this option was considered until many years afterwards. Although living animals could be protected, the Geological Survey perhaps could not see an obvious mechanism to protect fossils. Besides, national parks were the responsibility of the new Parks Branch of the Ministry of the Interior, created in 1911, whereas dinosaurs were clearly the responsibility of the Geological Survey, in the Department of Mines, created in 1907.

It is possible to assess Brown's likely response to opposition. In later years, he collected on the Greek island of Samos, and was refused an export permit for his fossil finds. "Barnum went up in a blaze of righteous indignation. . . . he immediately set out for Athens to do battle and settle matters himself," reported his second wife. "The man from Kansas was in a fighting mood." Barnum used technical arguments, persuasion by influential contacts, and bribery of officials to get his permit, only to find his wife had already smuggled the fossils onto the boat in case he was unsuccessful. A refusal from Canadian officials might have started a real bone war.

"THE PROPER PROCEDURE"

Reginald Walter Brock, an experienced geologist and director of the Geological Survey, solved the problem (no doubt in consultation with Lawrence Lambe). As dinosaur expert and palaeontological historian Edwin Colbert said, "quite correctly, he adopted the enlightened view that the proper procedure was to compete with Brown rather than to prohibit him from working." But who was available to collect dinosaurs for Canadian museums?

All of the experienced people in the Survey had died or moved on to other things, leaving only Lambe, by now in his late forties, and soon to become chief of the Survey's Palaeontology Division. Even if Lambe had wanted to go west again himself, he was getting the new museum's Vertebrate Palaeontology gallery ready for an opening, which eventually happened early in 1913.

Even if no Canadians with expertise were available, a rival team could be sent into the field to compete with Brown if the Survey hired experienced collectors from the U.S. And the Sternbergs were already selling American fossils to the Survey. This remarkable family by now included two generations of fossil collectors who worked together to supply fossils of all kinds from the American west. They were well known to museums in North America and Europe, and wrote regularly to potential clients to offer new discoveries for sale. At some point they had connected with the Survey, which was buying material from the U.S. for their new museum.

With hindsight, there is no doubt that Brown did science — and Canada — a favour by finding and collecting spectacular material. Dinosaur fossils erode away if they are not found and collected. Brown's finds certainly ended up in New York City and were lost to Canadian museums, but they would otherwise have ended up as mud at the bottom of the Red Deer River. By renewing interest in the Red Deer dinosaurs, Brown secured the future of Canadian dinosaur palaeontology.

THE BEST-KNOWN COLLECTOR IN THE WORLD

The father of the remarkable Sternberg family was Charles Hazelius Sternberg (1850–1943), who Brock called (not without reason) "perhaps the best-known collector in the world." His family moved to Kansas as a child, and Sternberg took an early interest in fossils, attended the University of Kansas, and began collecting for Cope in 1876. He accompanied Cope on a trip to Montana Territory in the same year, where they developed a rice-paste technique for collecting fossils. Sternberg devoted most of his life to his obsession with collecting large fossils, selling his finds to interested museums. He loved to be in the field, working in the service of science, but he was also a very religious man, who felt that by collecting and presenting fossils to the public he was doing God's work.

Sternberg married Anna Martin Reynolds in 1880, and three sons reached adulthood: George Fryer (1883–1969), Charles ("Charlie") Mortram (1885–1981), and Levi (1894–1976). George assisted his father in the field from the age of nine, and later Levi joined the team. At first, Charlie stayed at home to help his mother look after the farm. His education went as far as high school, then he went on his first collecting expedition at the age of twenty, and became as firmly hooked as the rest of the family. A sister, Maud, visited the field camps with her mother, perhaps cooking for the crew, but died while still a young woman.

By the end of the century, the Sternbergs had sold fossils to museums all over the eastern United States and Europe. They included two famous "dinosaur mummies," duckbills that had died and dried in a desert climate, preserving not just the bones but the shape of the body parts, all of the skin, and some of the stomach contents.

Sternberg was physically slight, lame, and deaf in one ear, handicaps that had not hindered his active life. He kept a firm hand on the family business — well into manhood his sons never knew anything about the financial dealings connected with the fossils they found and prepared. Sternberg wrote many accounts of his collecting, including two books, and received an honorary degree from Midland College, Kansas.

All of the male family members spent most of their lives collecting fossil vertebrates, and all played an important role in Canadian as well as American palaeontology. A grandson of Sternberg, Ray Martin, also described a fossil vertebrate from Dinosaur Park, though his career moved away from palaeontology.

Charles H. Sternberg.

WOULD YOU LIKE TO COME HERE?

At sixty-one, Charles Sternberg may have seemed elderly for strenous fieldwork, so Brock first approached his son Charlie Sternberg. "Would I come up to Canada and take a position and collect? I replied that father needed me and I didn't think it would be fair to leave him." The Survey then tried to purchase one of the Sternbergs' dinosaur mummies, but it already had been promised elsewhere. However, the Survey did acquire other fossils, including in September 1911 a large mammal, a *Titanotherium.* That winter, the Sternbergs agreed to mount the specimen in the new museum in Ottawa. Charlie recalls that "Brock was the Director, and Brock was a real go-getter. The result was 'Would you like to come here? 'What do you want?' 'Oh, about $100 a month. . . .' It was a silly amount. . . ."

Charles and George Sternberg left their home in Lawrence, Kansas, in March 1912 to make a tour of eastern museums, where Charles was able to see many of his own fossils proudly displayed. Arriving in Ottawa, "we found that the great room that was to be the exhibition room of vertebrate fossils, was filled with boxes and barrels, and there was not a tool in sight." With great ingenuity, the two Sternbergs worked out the methods of preparing an open mount, which they had never before attempted.

> At last our skeleton was mounted, but I notified Dr. Brock, the Director, and Mr. Lambe, the Palaeontologist too soon, forgetting the base had to be made of plaster. Just at the moment our plaster was hardening, and we needed all our wits about us, we ourselves being covered to the eyes with it, these gentlemen stepped down to view our mount. We were kept too busy . . . to entertain company.

George had already signed up for fieldwork in Alberta — with Barnum Brown's party. In later years George wrote, "after nearly seventeen years spent under the guiding hand of my father in the fossil fields of the West, there came an opportunity to go to New York and be associated with the American Museum; also the opportunity to visit new fields and learn the methods of some of the best preparators in the world. I felt there was much to learn so I accepted the offer made to me in New York." At the time it is possible that neither he nor his father had any idea that the rest of the family would go to Alberta in the same year for a different client.

"WE CROSSED THE INTERNATIONAL LINE"

When Charles found himself appointed the Geological Survey's Head Collector and Preparator of Vertebrate Fossils, he returned to Charlie's ranch in Wyoming on July 18 to pick up his collecting outfit, his son Levi, and his associate A.E. Easton. The next day the whole party — collectors, horses, and outfit — was Alberta-bound by train. At Calgary, "the metropolis of Alberta," Charles stopped to get a rowboat made while Charlie went on ahead. After eight days "our car arrived," and horses and men set out. Even before they got to the Red Deer River, they met eager myriads of mosquitoes.

> We were obliged to wear nets while traveling, and to keep a smudge going to protect ourselves and our horses from their murderous attack while we made camp.

They arrived at Drumheller, "a small town at that time, with a couple of stores," and camped three-quarters of a mile upstream. Despite a misprint in Sternberg's book, other evidence suggests it was about July 29.

Charles Sternberg, unjaded despite years in the wilds, looked with a fresh eye at this new and fascinating area.

> The valley of the Red Deer at Drumheller is a great chasm four hundred feet deep, cut by the river into the heart of the prairie. Across from plain to plain it is nearly two miles. Tributary creeks and coulees have cut narrow trenches back into the plain, while in the main valley, especially near the brink of the prairie, are long ridges, tablelands, buttes and knolls, pinnacles and towers. . . . All this region . . . has been transformed by nature's sculpturing into fantastic bad-land scenery."

The badlands did not make easy travelling, and the modern dinosaur hunter will have no trouble recognizing Sternberg's problems.

> ". . . the faces of the bluffs are covered with cherty chips that . . . slip under the feet and make it difficult to climb the steeper ascents. More than once I have measured my length on the steep surface, cutting face and hands by the impact. But, strange to say, when it was wet, and the clay beds were as treacherous as though covered with soft soap, wherever the cherty fragments accumulated,

> one could climb them in safety, as they were pressed into the slick clay, and held the feet securely as though there were spikes in our shoes.

The team was joined by a local man, Jack McGee, who had been a lumberjack and had a cabin outside the town. They searched the coulees west of Drumheller for fossils, but had only a rowboat to use on the river.

LIKE A DEAD DOG

The first bones found were of a "*Trachodon,* or duck-billed saurian. We soon began to find great numbers of loose bones piled up. . . . The best localities were above the river near the prairie level." Soon there was larger game.

> The Red Letter Day for us . . . was when Charlie found, on the 13th of August 1912, six miles west of Drumheller, the wonderfully complete skeleton of a duck-billed dinosaur. . . . It measured thirty-two feet in length. . . . The entire skeleton except the tail was present. Lying on its right side, the hind limbs were doubled on themselves . . . and the head bent towards the front limbs. The animal lay like a dead dog. I thought I had never seen anything so pitiful and forlorn.

Sternberg gives a more detailed description than Barnum Brown of the process of excavation. It suggests that the Sternbergs had less experience than Brown, and used a more improvisational approach, with things going wrong and a need to develop improvements on the spot. However, it is possible that Brown's account merely glossed over similar difficulties with some skeletons, as every fossil is found in a different situation, and new problems are presented to even the most expert excavator every time an excavation is made.

> My whole party worked in what I call, for want of a better term, "a quarry." The first thing we did was to remove with pick and shovel, the loose sand and clay, and lay bare a floor in the cliff large enough so that we would have plenty of room and could work down around the skeleton. We first traced the lateral spines, so there would

> be no danger of digging into the bone from above. This work was done with a digger and crooked awl, and only the merest trace of the bones was developed. When bones were exposed, they were instantly filled with shellac. They fall to powder on exposure without this precaution. The dorsal spines were traced in the same way and the ribs in front. Then we cut down several feet outside the skeleton so we could get under it. The skull was covered with burlap soaked in plaster and removed.

Unless he was taking the detail for granted, Sternberg does not seem to have been using an inner layer of tissue paper to stop plaster soaking into the bones as Brown did. Nor does he seem at first to have plastered the top parts of the entire skeleton before removing any pieces.

> The front limbs came next; and here we learned a lesson that was of inestimable value to us in taking up the vast bulk of the trunk region. When we turned the front limbs over, a lot of shattered rock fell out and threatened to bring the bones with it, and thus ruin the bones. No human being would have been able to mend these bones after they had been once jumbled together, so we . . . resolved not to attempt the big sections without covering the entire trunk beneath, as well as above, with plaster and burlap to hold the rock in place, and, of course, the broken bones, a surgical operation, in fact, by which the broken joints could be kept in place until they reached the skilled preparator in the Museum laboratory.

Sternberg's detailed descriptions of his method show that he was less sophisticated than Brown, but inventive and able to improve his methods as he went along. Unfortunately, once the skeleton was plastered their troubles were not over. They were trying to edge a large section of the skeleton up a runway of two-inch planks supported by a dirt platform, using a crowbar thrust into the ground to brace the rope.

> What was their surprise, when the section started in obedience to the law of gravity, to see the crowbar torn from Dan's hands and thrown to one side, and the section, unrestrained, gaining momentum at an amazing rate. The men below, who were guiding it, sprang out of

> the way, and the huge mass never stopped until it landed in the bottom of the wagon. The careful wrapping had prevented any damage, and without doubt it would have rolled to the bottom of the ravine without hurt.

BEANS AND BACON

Sternberg's team worked six days a week, and Charlie reminisced in later years about life in the camp.

> We had a lot of beans and bacon. Butter too. Marge is better in camp because it doesn't turn to grease in summer. We would dig a hole in the ground and put in some burlap and would put the butter in this lined hole. Had to be dried food — beans, potatoes and rice. Fresh meat when we went to town once or twice a month. We did not work on Sunday. I had a bath and washed my clothes. Always worn long underwear, even in the summer — it absorbed the moisture. I wrote letters and made my notes and maybe we would get visitors. If I had some specimens that had to be wrapped so that we could get away the next day, then I would work, but mainly it was washing.

The following winter, Charles and Charlie prepared their large duckbill.

> It was no easy undertaking to save and mount this wonderful complete skeleton; it was buried in fine sandy clay that was cracked in all directions, as were the bones — checked into thousands of fragments.

They were careful to "put a little life into the dead skeleton by straightening out the neck a little, and giving a sense of motion . . . to the tail." However, it was mounted in the position in which it was buried in the mud, with slight changes to reveal bones that would otherwise be unseen. Skin impressions were preserved over the pelvis, as well as the interlocking tendons that linked the muscles of the back and tail. Some parts of the skeleton presented great difficulty.

> . . . I was very doubtful whether it would be possible ever to mend the broken front limbs. They had been

> near the surface, and had been subject to the effects of frost, and plants, and their rootlets had severed the broken fragments and fed on their edges, destroying . . . the contact faces. However, Charlie's patience and endurance settled the question, and after six weeks of constant effort, he had filled the bones with shellac, picked up the fragments with small tweezers, cemented them, and pressed them into place. No one, without close inspection, could tell that the front limbs had ever been broken.

Once the new specimen was prepared, Lambe started to describe it. In a brief paper in 1913, he described the duckbill skeleton as *Trachodon marginatus* — "one of the most complete of the skeletons of Trachodon mounted in the museums of this continent" — and noted that it was "now being mounted in high relief preparatory to being placed on exhibition in the museum of the Geological Survey." After much confusion, this specimen eventually became the type of a new genus, *Edmontosaurus*.

RIVALRY AT STEVEVILLE

Until the summer of 1912, Barnum Brown had the Red Deer River to himself, and (even with George Sternberg in his team) perhaps had no idea that the rest of the Sternbergs would become his competitors on the same river. Although both the Sternbergs and Brown were from Kansas, they had spent years travelling in different areas, and seem not to have known each other except by reputation. There must have been mutual respect (Sternberg refers to Brown as "the greatest collector of extinct reptiles") but also a certain amount of envy of the other's discoveries. A flurry of letters greeted the news of the Sternbergs' arrival. Osborn wrote, "I all too deeply regret that the Ottawa museum which has laid dormant all these years should suddenly re-enter the field." Matthew agreed that "it's a pity, but nothing can be done but beat them to the best specimens."

Brown and the Sternbergs had somewhat different roles in the world of palaeontology. Although other people collected and sold fossils for a period of their lives, up to this time the Sternbergs were unique in their continuing career as freelancers in this field, selling to any museum that would buy their material. (Brown had purchased Sternberg material for his collections at the American Museum, and later published on some of it.) Although astute and knowledgeable, the Sternbergs did not have scientific training, and their material was largely studied by others. By contrast, Brown was an early (and still most remarkable) example of a collector who devoted his entire working life to one museum. Although an accomplished collector who preferred fieldwork, he had scientific training, and could describe his discoveries.

During three seasons, Brown had worked down the river towards Drumheller in a leisurely fashion, and he started his 1912 season in the same way. As long as he had the river to himself, he was in no hurry. But he had seen the rich Oldman Formation. Brown found the rival party close at hand when he arrived at Drumheller. His response was to abandon the Edmonton Formation and move straight down river to the Oldman exposures. He set up camp in Little Sandhill Creek, and found wonderful material, including skeletons of the duckbill *Corythosaurus* and the horned *Monoclonius.* The next year, in 1913, Brown's party continued to search the badlands around Little Sandhill Creek and found more material than he could well take out.

Sternberg's party was also back on the river. At Drumheller they purchased "a five-horse power motor boat; we also built a flat boat twelve feet by twenty-eight feet, upon the deck of which we pitched two tents, one for sleeping purposes, the other for a kitchen." In sixteen hours the motor boat towed the flat boat down the river, and they stopped at the little town of Steveville. "We were not far from Mr Brown's camp," noted Sternberg, and "Charlie and I went south to spy out the land."

Brown was annoyed to find his rivals on his heels again, and wrote to Osborn that he was "really provoked that they should follow our

The Sternbergs' floating camp.

The Sternbergs open a Chasmosaurus *quarry.*

footsteps so closely." At another point he wrote to Matthew: "Sternberg and his party are just below us and have taken out some fossils from our territory but we have at present no serious disputes. They have no regard for the ethics of bone digging." George Sternberg was still with the American Museum team, and must have found himself in a somewhat awkward position when his father and brothers arrived in the vicinity as competitors. George was under suspicion as a spy in the enemies' camp, and perhaps sought an opportunity to jump ship. At any rate, on his arrival at Steveville, Charles "was delighted to learn that my son, George . . . had been appointed on the Geological Survey of Canada, and would join my party." Brown complained to Osborn that this left him a man short, but at least the Sternbergs now would not be informed of his every move. However, relations remained cordial on the surface, and the parties visited each other's camps on days off.

In no time, Charlie had found "the skeleton of a carnivore . . . the most perfect one known to science at that time." He worked this find (later described as *Gorgosaurus)* with Jack McGee (who felt it was "altogether wonderful") while Charles, George, and Levi worked on other finds, including a *Corythosaurus* and other duckbills, and horned dinosaurs including *Chasmosaurus* and *Styracosaurus*. Lambe was able to make a brief visit to the Sternberg camp in September, and inspected all the active localities, "much to his delight."

If Lambe was delighted, Brown was ecstatic. Despite the competition, by the end of four seasons in Alberta, the American Museum expeditions had collected three hundred large cases, or three and a half railcar loads of fossils. Two-thirds of these were exhibition specimens, including twenty skulls and fourteen complete skeletons of large dinosaurs. Brown noted that these represented "many genera and species new to science, and defines the anatomy and distribution of several heretofore but partially known creatures."

CONFRONTATION ON SAND CREEK

In the 1914 season, the Sternbergs arrived on the Red Deer River on June 7, and recaulked the scow. George and Charlie towed it to make camp five kilometres (three miles) above Happy Jack Ferry, while Levi and Charles took the wagon.

In July Charles and Charlie spent a couple of weeks in Montana with a survey geologist, revisiting territory the elder Sternberg had visited with Cope more than forty years before. They then joined George and Levi, and on August 5 set up a new camp three kilometres (two miles) below Happy Jack. George found another *Chasmosaurus*, a skull of a duckbill *Prosaurolophus*, and a partial skeleton of an armoured dinosaur, *Euoplocephalus*. For a while, they had the assistance of an Oxford student, Patrick Disney, but when war broke out he returned home to enlist. On September 25, they dismantled the scow, and in high winds on September 28, loaded the last of their fossils at Denhart, a switch on the open prairie.

In June 1914 Brown's assistant Peter Kaisen arrived ahead of his chief. He tried to agree with the Sternbergs as to which territory each party would work, and felt that he had made a deal. At this time, vertebrate palaeontologists had little consensus on appropriate ethical conduct in competitive situations, so Sternberg sought guidance from Lambe. He was told that while it was clearly a breach of ethics to take up a skeleton already marked by rivals, no one party could reasonably lay claim to a large area, so that either party should feel free to prospect anywhere.

When Kaisen found the Sternberg party prospecting close to his camp, he felt the deal had been abandoned, and tried to confront them. Having lost his rivals in the rugged terrain, he went to their camp to express his annoyance. When Brown arrived in late July, he also expressed his frustration, at least to his superiors at home. He was particularly upset because George Sternberg had found a skull and partial skeleton of a horned dinosaur in an area Brown had already prospected — the Sternbergs had noted his footprints just 2.5 metres (eight feet) away. However, there was still a wealth of material for both parties, and the fine weather (with only three days lost to rain) allowed Brown to have what he considered his most successful summer, taking out eight fine skeletons.

There were lesser annoyances. Mice infested Brown's camp until a family of cats was acquired. One of the workmen brought in three orphaned bobcat kittens, which shared playtime with the domestic cats for some time. When the wild kittens started howling at night, the cook released them, but Brown lured one of them back with a fish. When the dinosaur bones were being packed for the museum, Rufus, the fast-growing and reluctant bobcat was (with difficulty) packed up for the New York Zoo, where he lived for some time.

If 1914 was exceptionally dry, 1915 had a remarkably wet summer. Brown wrote to his colleague W.D. Matthew: "It has rained all over this part of Canada as never before — the Red Deer River is out of its banks most of the time; ten feet of water was running over our last year's camp and came to where our fossils were parked last year. Mosquitoes are fearless of smoke, ferocious and in numbers equal to the Kaiser's army." The Kaiser was on their minds; the First World War had started in Europe the previous summer, with immediate involvement from Canada — the U.S. did not join the war until 1917, so the American collectors did not feel personally involved.

Perhaps to make sure the Sternbergs had not found something he had missed, Brown went at the end of the season to search the Milk River in Montana until "zero weather compelled cessation of operations."

The Geological Survey fielded two parties in 1915. George Sternberg was based independently in the Edmonton beds, and spent several months working from Big Valley down the sixty-five

Lowering a plastered block.

kilometres (forty miles) to Rowley, upstream from Drumheller where the Sternbergs had started work. He found a fine skull of the duckbill *Hypacrosaurus,* which Brown had described from a headless skeleton, as well as two skulls of a small duckbill later named *Cheneosaurus* (though it is now not distinguished from *Hypacrosaurus*). He completed the season by reconnoitring the badlands on the Battle River, further north, but found little of interest.

Charles, Charlie, and Levi were joined by a new assistant, Gustav Lindblad, who had just joined the Geological Survey. They began work on the Milk River valley, in southern Alberta, but did not find much dinosaur material. By the end of June they were in Dead Lodge Canyon again, and worked from there east to the Jenner ferry.

DOWNSIZING THE SURVEY

Brown returned to the American Museum early in January 1916, taking a railcar loaded with fossil dinosaur bones. There was a complete *Ceratops* and *Stephanosaurus,* a skull of *Monoclonius,* a skull and part of the skeleton of another armoured dinosaur, and the largest-yet discovered *Trachodon.* There was also a tiny but complete lower jaw of a Cretaceous marsupial mammal, and two tree trunks over twelve metres (forty feet) long. This brought to completion the museum's work on the Red Deer River, totalling over six years and four and a half carloads. Brown estimated it would take five technicians two years to prepare the material, and turned his attention to other fields.

Canada had entered the war in 1914. By 1915, the economy was in full swing, and there were shortages of many economic minerals and fuels. Although the Geological Survey's attention shifted to the economic and military applications of their work, funding for the dinosaur program continued through 1915. When Brown stopped work in Alberta, the Survey must have found it harder to plead priority for the program. A further complication came when, in February 1916, the central block of the Houses of Parliament burnt down, and the Victoria Memorial Museum was immediately put under requisition to serve as temporary quarters.

Funding for vertebrate palaeontology was under pressure, and four members of the Sternberg family were on contract. Charles was now sixty-six, and perhaps not the easiest man to deal with. His sons had shown themselves to be more than competent, and George was already working independently. "We were told that there was not enough money for field work and we would have to work in the lab," remembers Charlie. "But father said that if there was not enough money for fieldwork —

'There won't be any C.H. Sternberg either'." At any rate, Charles and Levi resigned from the Survey, while George and Charlie continued a much-reduced program. In the summer of 1916, George Sternberg continued his exploration of the Edmonton Formation while Charlie apparently prepared fossils in Ottawa. In the next season the brothers were in dispute about who should lead the field team. "So Lambe had to arbitrate and he gave me full charge of the field party. So George left. . . ." Charlie remained with the Survey for the rest of his long career.

It was perhaps during the downsizing period that George was given leave by the Survey to help with the local harvest. When a farmer asked about his regular work, he "explained that he went to western Canada to find and bring back prehistoric animals for the government museum. 'You must be a very brave man,' said his host, 'to capture such big, ferocious creatures.'

"MY FINEST DINOSAUR SUNK IN THE ATLANTIC"

When Charles Sternberg left the Survey in May 1916, he knew he could go elsewhere. He and Levi "soon afterwards went into the field for the British Museum of Natural History, London." Charles had sold fossils to the British Museum (Natural History) since before the turn of the century. Though Britain was in the midst of the First World War, the museum was not devoid of resources, and in May 1916 applied to the Percy Sladen Memorial Fund, administered by the Linnean Society, for funds "to employ a highly skilled and experienced collector Mr. Charles H. Sternberg." This would pay for "two months with his complete outfit (including at least two skilled assistants) . . . for the inclusive sum of $2000," and ". . . he would be willing to continue his work for the two following months at the same rate . . ."

In June the Fund's trustees agreed, but by this time the impatient Sternberg was already in the field with Levi. Sternberg's acceptance was reported to the fund in July, but by September 30 Sternberg reported (from Steveville in the first snowstorm of the season) that "the second two months ends today." Charles was worried "the weather is getting so bad we cannot mix plaster, Levi has all the ends of his fingers eaten off by the plaster making the work painful especially in cold weather." However, he was jubilant about the finds. "We have had the most wonderful success: three skeletons that can be mounted. But this last one in point of perfection far exceeds all the others."

The first shipment of fossils had gone, and despite the weather Sternberg expected to have the current shipment under way by the middle of October. Age was beginning to tell a little on Charles, who was "very anxious to get home as camp life does not agree with me in cold weather . . ." However, he was still planning and dreaming. "I would like to build up in the British Museum the third largest collection of Red Deer dinosaurs."

Charles obviously left Levi to attend to the details, as he next writes in November from Kansas, reporting that the second shipment (in twenty-two boxes) was on its way, insured for $2500 at a cost of $33.75. On November 7 Sternberg wrote again, indicating "that the Export Agent . . . expected to forward this shipment by the SS *Mount Temple* Nov 1st." On November 20 he wrote again after hearing of the safe arrival of the first shipment. Sternberg was not only anxious about his fossils, but funds were tight. "I was obliged to sell my home in Ottawa at a great loss, move my family here, and buy my old home at another expense. . . . I am naturally anxious to get some returns after my long and strenuous labor for your museum."

After Christmas Sternberg had a first cheque but no news. On January 19, the British & Foreign Marine Insurance Company queried whether the insurance "includes war risk," while about the same time, Dr. Smith Woodward told Sladen trustees of a rumour "that the ship carrying Mr. Sternberg's had been sunk." And indeed, a German U-boat had sent the *Mount Temple* to the bottom of the Atlantic.

Sternberg was horrified. "SHIPMENT LOST MOUNT TEMPLE COLLECT INSURANCE REMIT ANSWER STERNBERG," he cabled. Woodward hastened to pursue payment, but told Sternberg that the Sladen Fund refused to support further work. On February 9 Sternberg cabled again: "WHEN WILL YOU REMIT BILL SENT," and worried whether he should put the matter in the hands of lawyers. "I am not only paying interest on money that belongs to me but am prevented from going into the field for lack of it," he lamented.

Sternberg had to "mortgage my home to secure the money above what you sent me to the limit of my credit, and now, I have balanced my account there and find that I have only 163 Dollars with which to carry on my work this summer." Even in this state Charles went to Texas to join his son Levi for a month. By now the U.S. was in the war, and "Our boys in blue will help you clear the sea of those sea Pirates the scourge and curse of the world."

By May all is well, and Charles wrote from Texas to say he received "the full $2500 to day." The finances are lost in a welter of excitement at the new Texas finds, which are offered to the British Museum (Natural History). But Alberta is not forgotten. "I hope now to go into the Belly River after Dinosaurs, and in case I get a skeleton or some fine skulls, I will inform you. I am so sure of success I shall use every cent of the money from the policy and as much more as I can borrow. I will then offer the prepared material, Ward of Rochester has offered to buy all my material, but I will not sell any fine material to a dealer as long as the Museums of the world stand by my life work & support me."

Charles and Levi spent one more season in the Little Sandhill Creek badlands, finding a medium-sized *Gorgosaurus* (later named *sternbergi* after them), a skull and partial skeleton of an armoured dinosaur, *Palaeoscincus,* and a hooded duckbill. Charles tried in vain to persuade the British Museum to take his new finds as replacements, but they refused. The carnivore and armoured dinosaur went to the American Museum, and the duckbill went to San Diego. Sternberg moved to California for a while, and continued fieldwork into his late eighties. During the Second World War he returned to Canada to live with his son Levi in Toronto, where he died in 1943, at the age of ninety-three.

CUTLER AND THE CALGARY SYNDICATE

Although Brown and Sternberg were the principal figures in the Canadian dinosaur rush, there were others. The University of Toronto had an expedition in Alberta as early as 1908; since it was not very successful and the Royal Ontario Museum's serious work began in the 1920s, it more conveniently belongs in a later chapter. But in 1912 the career of the first Alberta-based collector got under way.

William Edmund Cutler was born in London, England, about 1878, and was brought to Canada as a child. He apparently acquired some scientific training, and perhaps worked in the field in Wyoming. By 1912 Cutler was homesteading near Dead Lodge Canyon on the Red Deer River. He explored the adjacent badlands, and by the time Barnum Brown came on the scene had already located a number of fossils. Cutler offered Brown a skeleton, in exchange for instruction in collecting techniques and a promise that the species would be named after him if it was a new one. Brown accepted the offer — the incomplete skeleton was of a horned dinosaur, with good skin impressions — and in due course named the new species *Monoclonius cutleri.* However, Brown and his crew did not get on well with Cutler, and they did not work together in subsequent years.

Cutler encouraged a group of Alberta businessmen to form the Calgary Syndicate for Prehistoric Research to fund his further work, the first Alberta-based dinosaur hunt. The Syndicate seems to have been organized through the Calgary Natural History Society, an active group supported by local businessmen. Its leader, Dr. Sisley, initiated the Syndicate on June 3, 1913, and James Davidson, president of the Crown Lumber Company, suggested getting twenty men to each subscribe $100.

Not much later in June, Cutler was already working in the field, as shown by a registered letter from Brooks, Alberta, dated June 16, 1913.

> I herewith tender you my resignation which will I presume take effect on July 16th or thereabouts because I have specimens being prepared to take another three weeks or more and then many boxes to make and then to bring it in to you and describe and enumerate contents. I hope under happier auspices & with more experience to be of greater value to you and your museum. . . . I have got you many specimens which as a new institution you ought to have and are of great value, scientifically, and also will awake a stronger interest in obtaining large grants for further work.

Dr. Sisley seems to have asked Brown to check up on their man in the badlands, for a month later (July 14), Brown reports:

> I have visited Mr. Cutler and find that he has a very good prospect . . . of a small dinosaur probably a Trachodon. This is a difficult specimen to handle, and I have frankly told him that it is beyond his present knowledge of preparation, especially if it continues into the bank of hard stone as now seems probable.

Brown liked to build local goodwill, and was also concerned that a potentially important fossil would escape him. His letter continues:

> I would be willing to prepare this for your society later in the season . . . and in case it proves to be a species not before represented in our collection would like to exchange a larger common skeleton for it . . ."

Despite Cutler's resignation, in October 1913 he submitted a bill to the Calgary society for four month's work at $125. Since a "fossil skeleton of a Trachadon [sic]" was added to the museum in 1913, together with a "fragmentary collection of ditto value $1000," the society perhaps felt that it had got its money's worth. This skeleton was partly prepared and displayed for years in the museum in Calgary. (It suffered some damage while on display, and was eventually exchanged with the National Museum of Canada for a *Corythosaurus* skeleton.)

In 1914, the secretary of the Calgary Natural History Society wrote to his M.P., R.B. Bennett, asking for support for further collecting. "I have seen many rare and valuable specimens exported . . . and it is a matter of deep regret . . . that these could not have been kept in our own country, where they rightly belong." The Minister of Mines consulted Dr. Brock, his deputy, who responded, "I do not know of any service that the society could render that would justify their services, which would enable us to make the society a money grant. . . . we would be very glad to let them have duplicate material . . . and allow our men to assist in labelling and displaying material. . . ."

Cutler was undaunted, and located a spectacular armoured dinosaur, a nearly complete skeleton with all of its armour intact. He may still have been collecting for the Syndicate (as a surviving list of fossils suggests), but he sold the specimen to the British Museum (Natural History).

Long the premier natural history museum in Britain, the BM(NH) had important connections with dinosaurs. Its first superintendent had been Richard Owen, who had been succeeded as vertebrate palaeontologist at the museum by his son-in-law, Arthur Smith Woodward (1844–1964). Its staff had a particular interest in "the colonies."

After the field season ended, Cutler yielded fossils to duty, and served with the Canadian Armed Forces in France. During the tedious hours in the trenches, Cutler must have thought of dinosaurs, for after the war he was back collecting in the west.

4

BIG GAME OF OTHER DAYS

"Today we must go to Africa for the biggest game; but there was a time in the dim distant past when America produced animals larger than any now living. . . . nothing remains of these creatures except their bones, and they are turned to stone. . . . their spoor has long since grown cold, and the hunt I shall describe is in consequence difficult."

BARNUM BROWN

AFTER THE WAR

Barnum Brown summed up his Alberta work in *National Geographic Magazine* in 1919. The United States had sent troops to France in 1917, and many of Brown's readers had not been able to spare a thought for dinosaurs. Brown himself had been away doing war work, advising the Treasury Department on taxation of oil properties.

When the armistice was signed in November 1918, the world heaved a sigh of relief. The "war to end war" was over. After the celebration, people everywhere turned their thoughts to the next priority in their lives. For a very small number of people that priority was Canadian dinosaurs.

The Canadian dinosaur rush had effectively ended when Brown shipped out his last boxcar full of fossils in 1915. Yet from this time on, somewhere in Canada someone was always working on dinosaurs.

Charles and Levi Sternberg had resigned from the Geological Survey, but they again became independent collectors, shipping material to the British Museum and others. George and Charlie Sternberg remained with the Survey, and each of them managed a single field season as the war continued. Cutler headed off to join the forces in Europe, but after the war he returned to the badlands. And other institutions — the Royal Ontario Museum and the University of Alberta — were planning how they could join the dinosaur quest.

In earlier years, the focus had been on fieldwork. Bits of dinosaurs were found, first by accident and as a by-product of surveys. Then specific expeditions were sent to find and collect them, and the collectors struggled with problems of functioning in the field, finding skulls and skeletons, and finding the best ways to collect, pack, and transport tons of fragile bones, often in a heavy matrix.

Fieldwork continued (especially as new institutions joined the search), but "collecting activity began to decline as the suspicion arose that the point of diminishing returns was approaching." The emphasis shifted from field to museum. Brown and the Sternbergs had between them shipped around thirty skeletons back to their respective museums, so work on existing collections became a priority. It was necessary to rescue the bones from their plaster jackets, so that they could be studied.

The most vital task facing the vertebrate palaeontologists was making sense of the trophies from the big game hunt into the past. Did the new specimens belong to species already known from elsewhere in the world, or from previous work in the same area? It was necessary to compare them with other specimens and published descriptions, and identify or name them and describe them for other scientists. Some institutions also hoped to involve the general public — to get the more spectacular bones on display and to write about them in non-technical terms.

All these activities involve the usual logistical issues of money, space, equipment, expertise, and (using what in those days was the appropriate term) manpower. With dinosaurs, those problems were present in spades. Compared with most of the fossil shells and corals museums had to deal with, dinosaur skeletons were huge, bulky, and awkward to handle. Every specimen needed bodies to move it, gallons of preservative, extensive space to store or display it, and a considerable acreage of text and illustrations to describe it adequately.

Nevertheless, during the 1920s and 1930s — between the two world wars — several institutions consolidated and developed their collections of Canadian dinosaurs. Some significant players — such as the American Museum of Natural History and the British Museum

(Natural History) — were outside the country. Others were in Ontario — the National Museum in Ottawa and the Royal Ontario Museum in Toronto. And in the west, the Calgary museum and the universities of Alberta and Manitoba continued or began to establish collections and exhibits in the areas where the dinosaurs were found. It is intriguing how the different institutions handled their dinosaurs — and how central the Sternberg family became to the Canadian effort.

NOTHING REMAINS EXCEPT THEIR BONES

Brown had shipped at least a dozen complete skeletons back to New York, together with isolated skulls and partial skeletons. In the lab a team of technicians was ready to cut open plaster jackets and chisel off surplus rock until the bones were clearly exposed. Loose pieces were glued together, and the bones hardened with shellac or other chemicals.

Brown was both collector and palaeontologist; while excavating he had a chance to think about the specimens, and correspond with other palaeontologists in the museum. While the fossils were being prepared for study, he started on identifications, surveying the large comparative collection in the museum, and published accounts of related dinosaurs.

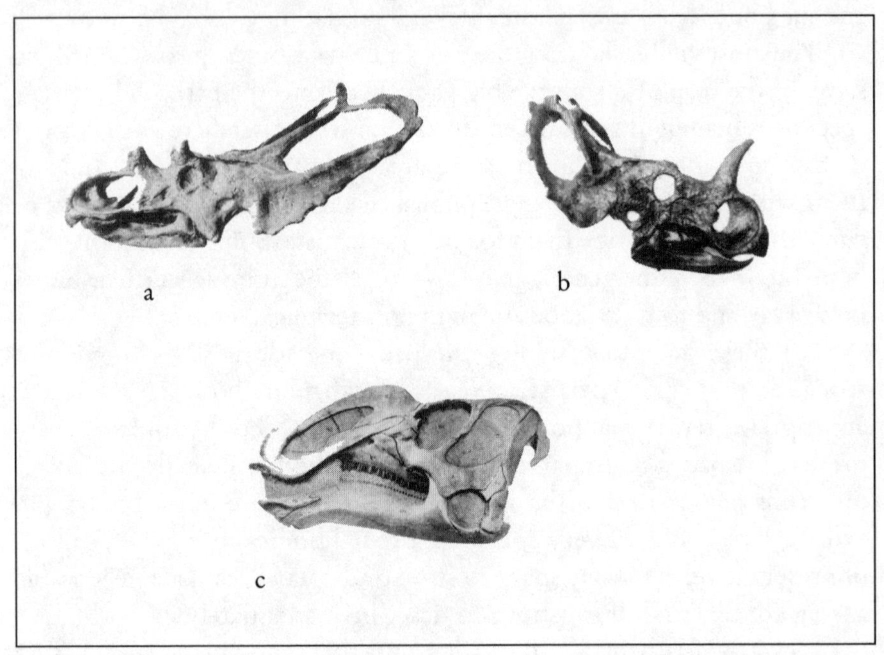

Skulls of a) Chasmosaurus *b)* Centrosaurus *and c)* Gryposaurus.

Only a few Canadian genera and species had been described during the years before the great dinosaur rush, some by Cope and Osborn, others by Lawrence Lambe. However, it was not clear at first that the incomplete Late Cretaceous material being found in western Canada was different from similar species already described from the American Jurassic. Often these fossils were fitted awkwardly into existing genera such as *Laelaps, Palaeoscincus,* and *Trachodon.* At first there seemed to be one kind of large carnivore, one duckbill, one armoured dinosaur and one horned dinosaur, then new and different material was attributed to different species within these genera.

Once Brown started finding complete skeletons, it became clear that almost everything being found in western Canada was new to science; only a few genera established from specimens from the American Midwest were of around the same age and closely similar. There were in fact several genera of duckbills, horned dinosaurs, and carnivores, and these groups must represent families.

Brown began publishing on his Alberta collections as early as 1912, and commented that the collection would be the subject of a monograph as soon as the material was completely prepared. Instead he published most of his scientific descriptions in separate papers in the *Bulletin of the American Museum* (which could be published quickly to ensure priority), but he also wrote for the publications of the New York Academy of Science, the Zoological Society of London, and even the Canadian Royal Society. Some of his material was also described by Osborn and William Diller Matthew (1871–1930), colleagues at the American Museum.

Without skulls, most duckbill and horned dinosaur skeletons appear much alike. However, when skulls were collected (with or without the rest of the skeleton), the differences were immediately apparent. Some of the duckbills were without crests, while others had a variety of elaborate constructions on top of their heads. Horned dinosaurs had huge carapaces stretching out behind the skull, with varying patterns of holes and horns. Preparators concentrated on the skulls, and Brown often published descriptions of them first, establishing his new names firmly in the literature; supplementary papers were published when the rest of the skeleton became available. He also kept careful track of the stratigraphy, and was able to show related dinosaurs from older rocks from Dead Lodge Canyon and younger beds (Edmonton Formation), giving the first indication of direct evolutionary relationships in dinosaurs.

DESCRIBING BROWN'S COLLECTIONS

Brown's first description was of the crested duckbill *Saurolophus osborni* found near the Tolman Ferry — the first complete dinosaur found in Canada. It was collected in 1911, and prepared and mounted in the winters of 1912 and 1913. Several other genera with crests were collected, including two skeletons of *Corythosaurus, Hypacrosaurus, Lambeosaurus,* and an incomplete skull of *Prosaurolophus.* Matthew later described another genus, *Procheneosaurus.*

Although armoured dinosaurs were not abundant (the best being the partial *Ankylosaurus*), horned dinosaurs were richly represented in the collection. The *Centrosaurus* found in 1914 had more of its bones than any other skeleton Brown ever found, for it was "complete in all details from the tip of the tail to the end of the nose with most of the bones articulated in position." Even the tongue bones were present! There were also skeletons of *Anchiceratops, Leptoceratops, Monoclonius,* and *Styracosaurus.* Another was initially called *Ceratops,* but moved to another genus, *Chasmosaurus,* in 1933, when Brown took the opportunity to name the species *kaiseni* after Peter Kaisen, "my friend and able assistant during many expeditions."

Not all of the treasure was buried in Alberta — some of it got lost in the American Museum's basement. Brown's assistant Roland Thaxter Bird (1899–1978) describes in his memoirs being asked to work on a *Monoclonius* skull, at a time when the museum was planning "to dispose

Barnum Brown admires Corythosaurus.

of some of its surplus." He and his colleagues were asked to fit together some pieces, as "Brown wants to know if there's enough of the skull here to be worth salvaging." As they worked, they found horns where there should have been a smooth frill. Brown came to examine the fossil and said: "This is no *Monoclonius* . . . we've got a *Styracosaurus* here. . . . This is a find. The only whole *Styracosaurus* skull in the world is in the Canadian National Museum in Ottawa." The next three days were spent piecing the skull together, and this led to another paper for Brown, and another new species. Brown's collaborator on this (and other projects) was Erich Maren Schlaijker (1905–1972), a palaeontologist and geologist who taught as Assistant Professor of Geology and Palaeontology at Brooklyn College between 1936 and 1950.

Before starting work in Alberta, Brown had collected two skeletons of *Tyrannosaurus* in Montana. Several large Canadian carnivores were collected — enough that exchanges could be made, such as the one that was sent to the Smithsonian. A particularly complete skeleton was added when one collected by Charles Sternberg in 1917 was purchased by the American Museum in 1918 for $2000; this at a time when a farm labourer earned $2 a day. This "most complete advanced carnosaur skeleton known from North America" was mounted by Peter Kaisen and Carl Sorensen and unveiled in 1921. In 1923 Brown and Matthew described it as the type of a new genus and species, *Gorgosaurus sternbergi.*

Osborn was anxious to add smaller carnivores to the collection, for their bones are fragile and not often found. He wrote to Brown in July 1913: "I do hope you have the good fortune to run across our long and much desired Ornithomimus: that seems to me now one of our greatest desiderata." And, indeed, in 1914 Brown found "one of the gems of the Museum's dinosaur collection," a nearly complete skeleton of *Struthiomimus altus.*

Brown has the reputation of having collected more dinosaurs than anyone else, but he did not return to Canada as a palaeontologist. However, he did prospect for oil for the Sinclair Oil Company, spending time in western Canada with Bird in 1939 before the Second World War, when they took the opportunity to make "the acquaintance of the Canadian dinosaur contingent." He retired in 1942, but continued working, and died one week short of his ninetieth birthday celebration, while he was planning yet another dinosaur expedition. In the words of the dinosaur hunter's first historian, Edwin Colbert, "Probably no other single individual has ever collected as many dinosaurs as did Barnum Brown."

"MY EO-CERATOPS SKELETON"

When Cutler, the west's first independent dinosaur collector, returned to Canada from the war in 1919, he began collecting for the University of Manitoba. Back in the Alberta badlands, he found part of a horned dinosaur, and worked on it alone late into the season, until at length he fell ill in the field. When he failed to come for supplies, a storekeeper in Steveville eventually went to check up on him, and brought him to the village to recover.

By 1920 Cutler was again in correspondence with Dr. Arthur Smith Woodward of the British Museum (Natural History), and in 1921 he was offering to "sell the collections of 1919–20 of Dinosaurian remains" for $2000. The collection included "a partial Eo-Ceratopsian skeleton . . . a good skull of Euplocephalus [sic] tutus (Lambe)" and a Sequoia (redwood) tree trunk. Cutler was investigating the possibility of taking palaeontological courses in London, and offered to collect in the field again for the British Museum (Natural History) in exchange for contributions towards his fees.

By February 1922, Cutler had arranged to collect in Canada for Woodward, and wrote again from Coleman, Alberta. Woodward had been asking if more dinosaurs were to be found, and Cutler cited "as proof positive, the returns of the Ottawa, Edmonton and Toronto expeditions. Of course," he continued, "one needs nowadays, after such thorough combing, to expect maybe one month or so of prospecting before receiving definite returns of finds of value."

On August 23 his real concern was cash: "By the way, I sent a small Ankylosaurus skull to a dealer who promised a lump payment of $200, which he now will not pay, and as I had need of money I was badly placed. Could I send it to you and trust to your endeavours to forward . . . as much as you can allow me for it?"

In 1922, Cutler secured a junior position with the Department of Geology at the University of Manitoba. He gave popular lectures on dinosaurs and became known as "Professor Cutler." He continued in correspondence with specialists at the British Museum and elsewhere.

> Would you please inform me whether you could interest the Trustees of the Museum to purchase my Eo-Ceratops skeleton in some 12 or so, medium sized boxes and amounting to perhaps two tons (or less) in weight. I am needing some £50 very badly for storage costs etc., and I will sell the specimen . . . for £125 . . . this is less than the

> outlay of the obtaining of this specimen, and I do not believe there is as good a skull and remains anywhere.

To justify his price, Cutler commented, "Prof. W.D. Matthew, Curator, New York Museum, has just decided that my skull of Ankylosaurus is that of Panoplosaurus . . . I offered it them for £25 . . . but they state that their Asiatic expedition had depleted their funds." Woodward purchased the specimen, and by May 1923 Cutler noted that it was sent. However, he was anxious for £50 in advance, as "I am greatly in debt to my storage merchant and indeed had great difficulty in inducing them to ship without an initial payment." There was clearly not much work available for a fossil collector: ". . . employment is difficult to obtain here now, precarious in holding and poor in remuneration, this coming autumn I have a small position offered me at the University but I wish to spend two months in the field this summer."

However, Cutler failed to get any further support for Alberta collections from the British Museum. The armoured dinosaur skeleton arrived there in 1915, but nothing could be done with it until the war was over in 1918. "So soon as the preparator, Mr. L.E. Parsons, could be spared from military duties, he started to remove the plaster and all superfluous rock, thus exposing the actual bones and skin. This task, which has occupied a large part of the last ten years, might even yet be unfinished had not the Geological Department acquired a pneumatic hammer for chiselling away the rock."

The British Museum had few vertebrate palaeontologists, and the specimen was described by the colourful Baron Nopcsa (1877–1933) in 1928. He described Cutler's skeleton under the name *Scolosaurus cutleri* ("Cutler's thorn reptile"). It was "the finest armoured Dinosaur ever discovered and the single one in which all parts of the dermal armour are preserved in situ." He suggested that the "living tank" had lived in semi-desert areas, and somewhat improbably speculated that it lived largely on grasshoppers. Later work has stressed its similarity to other armoured dinosaurs, and it is now included under Lambe's earlier genus *Euoplocephalus.*

Western Canada's first attempt to build a dinosaur collection had produced very indifferent results, and its collector's limited success largely benefited yet another out-of-the country institution. The next western attempt was to be more successful.

BARON NOPCSA

Ferenc Nopcsa von Felso-Szilvas (1877–1933) was one of the most eccentric characters in a field noted for strong personalities. The baron (whose name is pronounced "nopsha") was a Transylvanian aristocrat of considerable education, who wrote his scientific papers in four languages.

Nopcsa became interested in dinosaurs as an adolescent, studied geology and zoology at the University of Vienna, and published his first dinosaur paper in 1895. For the rest of his life he counted dinosaurs among his many interests, which included politics, geology, and archaeology, and the then-Turkish province of Albania, of which he planned to become king. He served in the Austro-Hungarian army and as president of the Royal Geological Survey of Hungary, and wrote many papers on dinosaurs and other fossils.

Alas, the baron came to a sad end. After alienating his colleagues in the Hungarian Survey by his attempts at reform, he took off on a 4800-kilometre (3000-mile) trip on his motorcycle with Bajazid, his Albanian male secretary and lover, on the pillion. In 1933, Nopcsa drugged and shot his friend, and then himself.

DINOSAURIAN BATTERING RAMS

John Andrew Allan (1884–1955), was born in Quebec and took two science degrees at McGill and a Ph.D. at M.I.T. After some fieldwork with the Geological Survey of Canada he became the first professor of geology at the University of Alberta in 1912. His vision for the department included collections for teaching and display, and naturally he kept a close eye on the activities of Brown and the Sternbergs over the next few years as they collected important fossils in his backyard.

When he heard that George Sternberg had returned to collect at Little Sandhill Creek on speculation in 1920, Allan sought funds. It must not have been easy; a few years later (while naming a fossil turtle) a researcher explained it was "in honor of Dr Allan, who in the past two years, under difficult circumstances, has so successfully brought together a collection of vertebrate fossils." Nevertheless, he found the money to purchase Sternberg's collection, and hired him to work through the winter to prepare the material. In spring 1921 George returned to the field and collected a fine *Corythosaurus* and a carnivore skull. He also found a beautiful skull and partial skeleton of another dinosaur.

> I saw three small teeth glistening in the sunlight. . . . They were small but perfectly preserved. I soon had a floor laid bare and in less than an hour I knew I had a perfect skull of this little animal as well as most of the skeleton. . . . I went back to camp taking the skull with me. I dug up every bit of literature dealing with it that I possessed. . . . I wrote a long letter to the University of Alberta, telling them all about my find. I was thrilled to the very marrow to think that I at last had been the fortunate discoverer of this little animal.

This little dinosaur solved a problem that had been puzzling other collectors for years. In 1902 Lambe had picked up a solid piece of bone that appeared to be a thickened skull cap of a small dinosaur, which he named *Stegoceras.* George's skeleton showed what the little dinosaur was like, and speculation suggested that the males used the thickened skulls as battering rams, perhaps crashing their heads together in macho competition like bighorn sheep.

In 1922, George Sternberg joined the Field Museum of Natural History in Chicago, and collected for them in his last season on the Red Deer River. Except for a few months at the University of Alberta in 1935, when he finished preparing the material he had collected, George did no more work in Canada. He eventually settled in his home state of

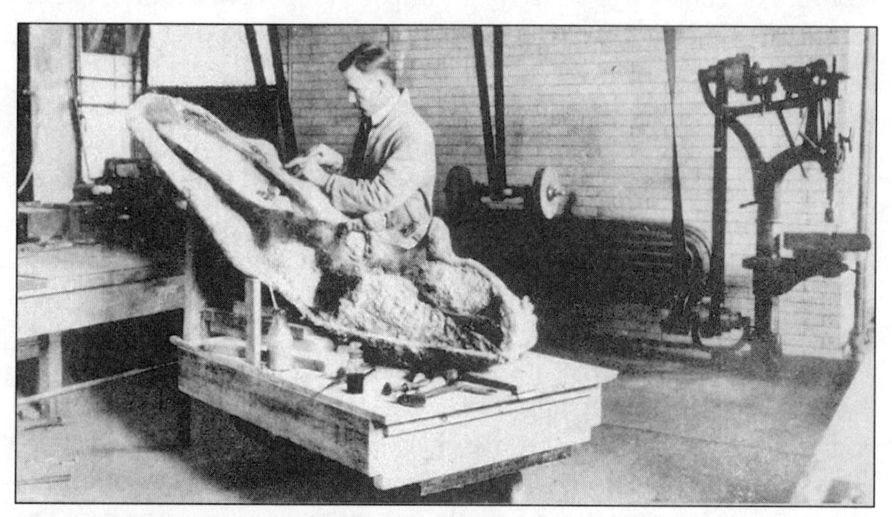

George Sternberg prepares Chasmosaurus.

Kansas, where he built up a museum that became the Sternberg Memorial Museum in Hays.

Allan was not a vertebrate palaeontologist, so he invited an American palaeontologist, Charles Whitney Gilmore (1874–1945), to study the material. As a child in New York state, Gilmore had been taken to Ward's Natural Science Establishment and come home enthralled. He studied at the University of Wyoming, and (after a few years at the Carnegie Museum in Pittsburgh) joined the staff of the United States National Museum in Washington, D.C. Here he rose to be a curator, and worked extensively on fossil reptiles. He described a number of new species from Alberta. He examined George's *Stegoceras* material, decided that its teeth were the same as the little-known *Troodon,* and used that name for the skeleton — creating a problem that was not sorted out until recently. He also began the systematic review of species already described, noting that the hadrosaurs were "in a somewhat chaotic state."

ROM WASN'T BUILT IN A DAY

"They had mounted the Mosasaur skeleton I had sold to Professor Parks some years before. It was the only large vertebrate on exhibition."

Toronto's Royal Ontario Museum (ROM) had opened in March 1914, and Charles Sternberg, travelling west in June, had stopped off to visit. Though Sternberg found the gallery disappointing, University of Toronto scientists had already begun collecting fossil reptiles, and in later years Canadian dinosaurs were to become an important feature of the ROM's collections, research, and exhibits.

There had been interest in a museum in Toronto as early as 1853. Collections were being made, particularly under the Professor of Natural History at University College, William Hincks (Huxley's successful rival), who "had become the driving influence to establish a biological museum in Toronto." Various ad hoc collections were established in temporary quarters in the university.

Leadership was later provided by Byron Edmond Walker (1848–1924), who started life as a farm boy, collected fossils with his father, and became a leading banker. He joined the Canadian Bank of Commerce, and spent several years in its branch in New York, where he became familiar with some major museums. In 1892 he was appointed to the board of trustees of the recently united University of Toronto, and when in 1899 he gave a presidential address to the Canadian Institute he chose the topic "Canadian Surveys and Museums and the Need of Increased Expenditure Thereon." In 1905 Walker made sure that his cherished plan for a museum was included in the recommendations of a Royal Commission inquiring into the university. On the faculty, Professor William Arthur Parks (1868–1936) "from the time of his initial appointment until the time of his death forty-three years later, worked assiduously for the establishment of the museum and for its growth and development."

By now, the academic community was well aware that dinosaurs were to be found in the west. In 1908 the "amiable, cheerful and capable" Benjamin Arthur Bensley (1875–1934), Professor of Zoology, set out for Alberta. At Berry Creek on the Red Deer River he made a collection of "more or less fragmentary" material, including an incomplete duck-billed dinosaur skull.

In 1913 several independent museums were created, each to share the same building. Each was managed by a university professor, working without extra pay. Bensley became director of the Royal Ontario Museum of Natural History (soon renamed Zoology), while Parks was in charge of palaeontology. In 1912 Parks sent Professor of Geology Alexander McLean to Munson Ferry, to collect in the Edmonton Formation. As with so many earlier collectors, McLean's inexperience prevented him from bringing back more than "a large collection of dinosaur bones . . . but no skeleton or head."

Parks had received the first Ph.D. in geology ever awarded in Canada. He added palaeontology to the curriculum at the University of Toronto, and did summer fieldwork with the Geological Survey. Initially he worked on invertebrate fossils, but recognized the importance of dinosaurs for the new museum, and during the war "longed for the time when financial assistance could be obtained whereby regular expeditions might be sent to the Red Deer." He made his first expedition to the Red Deer River in 1918, at the age of fifty, with a grant secured by his friend Walker. He was assisted by Robert Wilson, and they found a skeleton of a duckbill later named *Kritosaurus.* Despite inexperience, they collected it successfully, but there was no preparator available.

Levi Sternberg was hired as head collector and preparator, just in time to go into the field with Parks the next year. The following winter he and Wilson prepared the new skeleton, and once it was on display Parks had no difficulty securing further funds. When Wilson left, Levi's former colleague from the survey, Gustav Lindblad, joined the museum in the fall of 1921. The two had close ties, as Levi had by now married Lindblad's sister, Annie. This time Levi led the field camps, while Parks came out to visit whenever he could. "It is interesting," comments Loris Russell "that the dignified and rather aloof professor of the campus became in the field the most congenial of campmates."

In 1922, another technician, Ralph Hornell, joined the staff, and in 1923 John Rickett was added. Year after year, the expeditions went out, mainly to Alberta but also to mammal sites in Saskatchewan and the U.S. Parks described their work in the museum:

> The removing of the bones from the matrix, their preparation and assembling, and the final mounting of the skeleton are done in the laboratory. The process is tedious and delicate and requires expert manipulation. As a bone is gradually exposed by carefully chipping away the matrix, it is repeatedly soaked with thin shellac. This material hardens in all cracks and interspaces, making possible the further removal of the matrix. . . . Without the binding effected by shellac, the removal of most bones would be difficult, and in many cases impossible. Many bones, thus prepared, are still incapable of supporting their own weight. They must be strengthened by wires passed through them or by bands of soft iron, forged to shape, and bolted to the bone.

Parks (despite becoming head of his department in 1922 and continuing his other research) described the material that came in, and made major contributions to Canadian dinosaur studies. From 1919 to 1935 he published more than twenty papers on dinosaurs (carnivores, ceratopsians, and hadrosaurs). Parks described a number of new genera and species. One was a particularly striking duckbill, *Parasaurolophus,* with a long crest; the type species was named *walkeri* after Edmund Walker, who was at the time chairman of the museum's board. Horned dinosaurs included *Arrhinoceratops* (still known from a single skull) and new material of *Centrosaurus,* as well as an armoured dinosaur, *Dyoplosaurus.* In later life, Parks received many honours, but would perhaps have valued most another dinosaur, *Styracosaurus parksi,* which was named in his honour after his death by Brown and Schlaijker.

In his younger days, Levi Sternberg had been the joker of the family, and remained "a congenial companion, always joking and teasing." He stayed with the department, leading field expeditions (which he always enjoyed most) until 1954. Unlike his older brothers, he was "indifferent to study and scientific investigation" and he stayed a chief technician. "Professor Parks was to say on more than one occasion 'Too bad Levi didn't go to college — he could have been head of the department after me.' This was said not in a condescending way, but with real regret." In later years, Levi provided a home for his father in Toronto, until Charles Sternberg died in 1943.

"A TROVE OF SPLENDID FOSSILS"

By 1916, the Sternbergs had shipped to Ottawa sixteen complete or partial dinosaur skeletons — later described by Peter Dodson as "a trove of splendid fossils." Charles Sternberg calculated that "it will take twenty years of careful labor by four competent preparators to get all the material in shape for study and exhibition." Some of Sternberg's photos show the lab — equipped to handle dinosaurs — that was fitted out in 1912 in the basement of the new museum. By modern standards it was a crude workroom, lined with brick and with unprotected equipment; at the time it was seen as a fine facility. An overhead trolley and hoisting block could lift two tons, and electric drills and delicate dental tools and brushes simplified the cleaning of fossils. A furnace allowed forging of iron supports, so that dinosaurs could be mounted as free-standing displays.

Once the Canadian dinosaur rush began, Lambe could work on material collected by the Sternbergs — specimens of much better quality than Lambe's earlier collections. However, he had little comparative material to work with, and belonged to a government organization, with the inevitable red tape (publications had to be approved by the director). Although the Geological Survey had its own series of publications, these appeared slowly, so that Lambe's work often appeared in print after long delays.

Barnum Brown started research in the field and had a large team of technicians and several collaborators who helped him get material into print. Since the two teams were collecting in the same areas, they inevitably acquired similar material. Where both Brown and Lambe described the same new genus, Brown's competitive advantage often allowed him to publish his description first, and so (by the internationally accepted rules) his names had priority. Consequently, a number of Lambe's names have had to be abandoned. This cannot have helped relations between them; it is not surprising that Lambe sniped gently at Brown and his work in some of his papers, while Brown ignored some of Lambe's genera.

Lambe did important work, particularly on a number of complete skeletons of duckbills. These were the basis of several new genera, including *Edmontosaurus, Gryposaurus, Stephanosaurus,* and *Cheneosaurus*

— "goose reptile" — "on account of the supposed resemblance of the specimen, when viewed in profile, to the outline of the head of a goose." Horned dinosaurs included *Chasmosaurus*, *Eoceratops*, and the many-horned *Styracosaurus.* Armoured dinosaurs were not neglected, for he described the new genus *Panoplosaurus.* A new family was created for the bone-headed *Stegoceras,* which Lambe placed provisionally with the armoured dinosaurs. He also continued his interest in the larger carnivores, giving a detailed description of "Charlie's carnivore," *Gorgosaurus.*

Sadly, Lambe died in his mid-fifties; had he lived he would no doubt have achieved much more. He was Canada's first vertebrate palaeontologist of any stature, had pioneered research on Canadian dinosaurs and other groups, and had lived to play a major part in the establishment of the Canadian Upper Cretaceous beds as one of the richest dinosaur deposits in the world. It is fitting that a striking duck-bill dinosaur genus, *Lambeosaurus,* was named after him by Parks.

"THIS BRIGHT YOUNG CANADIAN"

After the First World War, with Lambe recently dead and the Royal Ontario Museum getting serious about dinosaurs, there were job opportunities for an experienced Canadian vertebrate palaeontologist equipped to study the western dinosaurs. Could a qualified Canadian candidate have been found? William Diller Matthew (1871–1930) was a Canadian on the staff of the American Museum of Natural History in New York, and had already published on Alberta dinosaurs.

Matthew was the son of George Frederic Matthew (1837–1923), who was a Customs House Officer in New Brunswick, and a part-time geologist of considerable stature who published more than two hundred papers. George tried to initiate a geological survey in his province as early as 1864, and four years later worked for the Geological Survey under Logan. He gathered a large collection of fossils, which he curated for the New Brunswick Museum. Although he never had a chance to work on dinosaurs, he did do important work with Carboniferous fossil footprints at Joggins and other Maritime sites.

His son William naturally became interested in palaeontology (finding a giant trilobite at the age of six), and went to study first at the University of New Brunswick and then at the

Columbia School of Mines in New York City. Here he was taught palaeontology by Osborn, who recognized "the potential of this bright, young Canadian." In 1895 he joined the American Museum staff. Two years later his father attempted to interest him in a position at McGill, and William wrote that "I should like very well to be in an institution like McGill, and in my own country." However, he was paying his younger brother's way through Columbia, and so stayed on in New York — for thirty years. In due course he became curator of the Vertebrate Paleontology Department, and thus Barnum Brown's boss. He took a particular interest in dinosaurs, and wrote extensively on Alberta material, often jointly with Brown. In 1915 his 162-page book *Dinosaurs, with special reference to the American Museum collections,* was intended as a museum guide, but also was essentially the first popular book on dinosaurs ever published.

Matthew moved to California in 1927, as chairman of the new Department of Paleontology at the University of California, but died of a kidney disease in 1930. A memorial tribute hailed him as "one of the most influential vertebrate paleontologists of this century."

Matthew never lost his links with Canada, for some of his family remained there, and one of his brothers fought and died in a Canadian regiment in the First World War. Before his fiftieth birthday, Matthew was proposed for election to the National Academy of Science, the premier scientific organization in the United States. However, full membership was limited to U.S. nationals, and the proposers discovered that Matthew was still a Canadian citizen. He became an associate instead, and was also elected a fellow of the Royal Society in the U.K. When Matthew died, his ashes were shipped back to New Brunswick.

Matthew left a special legacy to dinosaur studies, for his daughter Margaret, who trained at the California College of Arts and Crafts in Oakland, came to the American Museum as a palaeontological illustrator in 1931. Two years later she became the wife of Edwin Colbert, by then on his way to become the leading dinosaur specialist of another generation, and also the first major historian of dinosaur research. Colbert eventually wrote a biography of his father-in-law, celebrating a Canadian vertebrate palaeontologist who wrote extensively about Canadian dinosaurs yet never had a palaeontological job in Canada.

CHARLIE

When Lambe died in 1919, he left an incomplete description of an armoured dinosaur that he called *Panoplosaurus* ("fully armoured reptile"). The completed text about the skull and armour was published under Lambe's name, but Dr. E.M. Kindle (who was head of the Survey's palaeontology department) encouraged Charlie Sternberg to describe the rest of the animal. After being reviewed by Gilmore in Washington, Charlie's first written contribution on dinosaurs was published in 1921.

Although Charlie (like his brothers) had only high school education, and had previously only written a couple of natural history notes for publication, he had become fascinated by dinosaurs while collecting with his father. "I got every publication that I could find describing the different ones," remembered Charlie. "I am a self-educated man because I got no training in palaeontology in my school. . . . But when I found a specimen I studied it in the field. I collected it, I prepared it . . . all this time I was studying."

Upon Lambe's sudden death, Charlie remembers, "there was no one to carry on." It created an opportunity that his brother Levi lacked at the more academic Royal Ontario Museum. Other candidates were considered:

> . . . they thought of Gilmore who was in the U.S. National, but he didn't have a Ph.D. but I was making this discovery and that discovery. My immediate chief Dr. Tindale . . . came to see me — and said that no one should be allowed to do scientific work without a Ph.D. and I had to stop doing my work. So I went to see the Director . . . he knew I had been working every night until six and doing research on my own time. But he said if I could reach the high standards, they would publish the findings . . . the door was opened and I walked in.

Charlie continued not only as a field collector, but also took over Lambe's scholarly role. He took out Canadian citizenship in 1922, and continued his career with the Survey and the National Museum of Canada (which separated from the Survey in 1948) that lasted for an astonishing thirty-eight years. After retirement in 1950 he continued as a research associate. In all, he published some sixty-five scientific and popular papers (of which forty-five — almost seventy percent — were on dinosaurs, in which he described seventeen new genera or species).

Charlie — remembered by Peter Dodson as "short, taciturn and doggedly descriptive" — had many responsibilities. He was for many years the only vertebrate palaeontologist with the Survey, though he later had assistants, including Harold Lowe and J. Skillen, and student assistants, including T. Potter Chamney and Loris Russell. "He was not well educated," remembers Wann Langston, Jr., "but he was an astute observer. He was absolutely truthful, trustworthy. He was well aware of his own limitations, and he never overstepped them with speculations about things he knew he didn't know anything about."

Charlie had to collect and study other vertebrate fossils. As well as such contemporaries of the dinosaurs as ichthyosaurs, crocodiles, lizards, and turtles, his scientific work also covered early fossil footprints in Nova Scotia, and more recent fossils, such as bison, mammoths, and seals. However, dinosaurs were his first love; at Lambe's death the museum had much dinosaur material that was undescribed, a constant reminder that new fossils were available to be discovered in western Canada whenever Charlie could get away into the field. A few examples will illustrate the range of his research.

Material already in the collection described by Charlie included a new genus of armoured dinosaur, *Edmontonia.* In 1935 he made a systematic review of all the hadrosaurs, preparing the skulls that were still in their field jackets, and described three new species. He also continued to collect new material, as a few examples of his later work will show.

In 1921 he returned to the area in southern Saskatchewan where Dawson had first found dinosaurs in Canada. There he found remains of carnivores, duckbills, and horned dinosaurs, and had them "all identified or verified" by Gilmore at the Smithsonian. The duckbill — "the first found in Saskatchewan of which a considerable part of the skeleton was preserved" — he later described as a new species of *Thespesius.* The horned dinosaurs belonged to *Triceratops*, a large and characteristic late, horned dinosaur. These dinosaurs showed that these beds were at the highest level of the Cretaceous (Lancian stage), younger than those studied so far in Alberta. Later, in 1946, Charlie found other deposits in the Scollard area of Alberta that were also Lance Formation.

In 1928 Charlie found bits of two new small carnivores, which he described in 1932 as *Macrophalangia* and *Stenonychosaurus*. Now that many complete dinosaur skeletons were available, it was clear that carnivores of any kind were relatively scarce. The small, slender ones had much more fragile bones and were even more infrequently found, so increasingly collectors began to look for them.

In 1937 Manyberries, in southeastern Alberta, was the subject of another trip. Here, Charlie recalls, "my ambition of twenty years was realized" when he discovered two skulls of the horned dinosaur *Monoclonius.* Charlie also collected a splendid *Lambeosaurus,* which (although it became well-known from a field photograph) remained in its packing cases for another thirty years, when I supervised its preparation.

In 1946 Charlie visited Scabby Butte (whose fossils had not been studied since McConnell's and Weston's time) to follow up finds made by geologist colleague O.A. Erdman. Here he found a dinosaur specimen weathered completely out of the rock and scattered over the hillside, and commented wryly that "much had been taken away by settlers who have used the spot as a picnic ground for many years." Although related to the horned dinosaurs, its huge bony skull was apparently without horns, and Sternberg named the "freakish development" *Pachyrhinosaurus.*

Another hornless ceratopsian find was made in 1947, when Sternberg and his student assistant T. Potter Chamney found no less than three skeletons of *Leptoceratops,* one being "the only absolutely complete ceratopsian skeleton known." Dodson describes Sternberg's work on these important skeletons, then adds, "The skeletons were exhibited at the National Museum of Canada — without fanfare or puffery. In consequence of this typically Canadian modesty, *Leptoceratops* has never enjoyed the renown that the quality of its fossils and its evolutionary significance naturally merit."

This brief summary gives only a glimpse of some of Sternberg's interests. Some of his papers are notable not just for the new forms named but for gentle probings into new ideas — matters of behaviour, ecology, and evolution — some of which will be discussed in later chapters.

For a short time, it seemed as though Charlie would be the middle generation of a palaeontological family. His son Raymond M. Sternberg (1912–1992) started a career in palaeontology, working with his father in the field. In 1940, from a lower jaw found in Alberta, Ray described what he considered to be a fossil bird, calling it *Caenagnathus.* After years of controversy, it is now generally accepted as a toothless dinosaur. During the Second World War Ray changed both his surname (adopting his mother's name, Martin) and his career. When I met him in the late 1980s he was living in Victoria, B.C. At that time, he was so far out of touch with palaeontology that he asked me, "Does my genus still exist?" I was able to assure him not only that it did, but that another species had been described thirty years later.

DINOSAUR TRACKS ALONG THE PEACE

"It is puzzling why . . . the . . . Red Deer River, which has yielded so many well-preserved skeletons of dinosaurs, has afforded only one specimen of tracks, whereas the Connecticut Valley, from which thousands of tracks have been collected, has yielded very few skeletal remains." Charlie Sternberg also commented that although ripple marks (suggesting sandy beds formed in shallow water) are common on the Red Deer River, tracks are extremely rare. Charlie did record four dinosaur tracks, found and collected by a Geological Survey party near Rumsey in Alberta in 1926. He considered them to resemble those of the little bird-mimic dinosaur *Ornithomimus*, and so named them *Ornithomimipus* ("ornithomimid-like foot").

Footprints already had been found by geologist Frank Harris McLearn (1885–1964) along the Peace River canyon in northeastern British Columbia. It was not until 1930 that Charlie was able to visit and study this then-remote site. He found a treasure trove of tracks in Lower Cretaceous rock, at least ten million years older than the Red Deer beds.

The Peace River is now drowned by dams, but at that time it fell 83 metres (272 feet) over 19 kilometres (12 miles), roaring through falls,

Peace River trackways.

rapids, and huge standing waves. Mackenzie and many subsequent travellers found it quite unnavigable. The tracks are found in gently dipping sandstones along the river shores. The sides of the canyon are steep and rocky, though locals had gouged trails through the forests on some of the gentler slopes to work the exposed coal seams — one as much as two metres (six feet) thick. One mining family, Neil Gething and his sons, had found more tracks since McLearn's visit, and they helped Sternberg to collect some tracks and make plaster casts of others.

These sandstones also had ripple marks, and there were mud cracks in the shales, suggesting to McLearn "deposits of shallow flood-plain lakes whose bottoms were exposed as mud flats at times of low water." Charlie found more than four hundred individual footprints, some in long trackways.

NAMING TRACKS

The Reverend Edward Hitchcock (1793–1864), whose studies of Connecticut trackways are mentioned by Sternberg, was a remarkable pioneer. He thought his tracks were made by birds, but he developed methods of describing and naming them that are still in use today.

It is impossible to know with certainty what animal made an individual footprint unless it has literally dropped dead in its tracks — so trackways cannot be conveniently referred to by the name of their makers. Yet palaeontologists recognize many different kinds of tracks, and need to be able to name them in some way so that they can talk about them. The solution is a parallel nomenclature, applied to tracks alone, in which they are given scientific names just as if they were animals. Where a correlation between a track and an animal seems certain, track names may be derived from the names of dinosaurs known from bones. Following this tradition, Sternberg named the Rumsey tracks *Ornithomimipus angustus* ("narrow *Ornithomimus* tracks").

However, more caution is usual, and Sternberg's later names are often based on such factors as location or appearance.

Sternberg described seven new genera, including *Amblydactylus* ("blunt toed"), *Columbosauripus* ("the foot of a saurian from British Columbia"), *Gypsichnites, Irenichnites* and *Irenisauripus* ("saurian or lizard foot of Peace River country"), and *Tetrapodosaurus.* McLearn and Gething were acknowledged in the names of species.

Bones always come from a dead dinosaur, but tracks have been made by living ones, and can give direct information about their behaviour. Charlie's interpretations give glimpses of prehistoric moments, of dinosaurs walking in very soft mud, changing the speed of their walk, or stopping altogether. In later years, these tracks would yield even more exciting information.

A PERIOD OF CONSOLIDATION

Late in life, Charlie Sternberg was once introduced as an expert on dinosaurs. "Oh no," he said. "If I had chance to do it all again; then I might be an expert." Despite his modesty, more than anyone else in the middle years of this century, Charlie was Canada's "Mr. Dinosaur." Already an experienced fossil collector when he came north in his late twenties, his working life in Canada bridged the period from the Canadian dinosaur rush through the long period of consolidation and new exploration between the wars. He received many honours — elected to the Royal Society of Canada in 1949, and honorary degrees from the University of Calgary (1960) and Carleton (1974). He remained the "grand old man" of Canadian dinosaur research on into the new renaissance. After his wife, Myrtle, died in 1977, Charlie continued to live in an apartment in Ottawa, still only a few blocks from the museum.

The National Film Board featured him in a film, *Charlie,* and I was able to arrange for an audiorecording to be made by the Provincial Archives of Alberta. I visited and talked with Charlie when he was in his nineties, and was delighted to learn that before his death in 1981 he had heard the news that Alberta was planning a major dinosaur museum.

The Canadian dinosaur rush filled a vacuum by bringing in outside expertise, though some of the most significant dinosaurs left the country. But Brown and the American Museum had shown what could be done with expertise, dollars, and enthusiasm, and the lesson eventually bore fruit in Canada.

During the period between the wars, Canada made notable progress in collecting and studying its own dinosaurs, as slowly but consistently, the two major natural history museums in central Canada followed Brown's example. Lambe's pioneer work was continued by two major vertebrate palaeontologists, as Parks and Charlie Sternberg presided at the Royal Ontario Museum and Geological Survey (National Museum). Toronto also gave a home and a career to Charlie's brother Levi, while the University of Alberta took further advantage of George's expertise before he returned to the United States. A group of capable technicians was trained, and among Charlie's student assistants were to be found some of the next generation of vertebrate palaeontologists.

By the outbreak of the Second World War many elements of the fauna of the Upper Cretaceous of Alberta had been collected and described. The hardcore four of Brown, Lambe, Parks, and Charlie Sternberg published between them around a hundred papers on Canadian dinosaurs. Three successive formations had revealed a huge and diverse fauna, with large and small carnivores, many duckbills and horned dinosaurs, and a growing number of armoured dinosaurs, as well as several imperfectly known and hard-to-classify oddballs. (Of the major groups of dinosaurs then known, only the great sauropods and the plated stegosaurs, now well-known from the Jurassic of the American Midwest, proved not to be present in Canada.)

Patient work had described many kinds under many names, and gradually attention was shifting from new discoveries to reviewing the material, sorting out a meaningful classification, and trying to achieve a better understanding of what had been discovered. But this knowledge was as yet largely in the hands of a few specialists.

Unless they lived on the prairies or within reach of Ottawa or Toronto, there was no reason for the public to take an interest in dinosaurs. *Arctosaurus* was forgotten, and since *Bathygnathus* was no longer a dinosaur, there was no maritime connection. Only the dinosaur tracks in B.C. gave promise that there might be finds in other parts of Canada — but dinosaur tracks were not taken very seriously in those days even by most palaeontologists.

"A WORK OF NATIONAL IMPORTANCE"

In a 1935 report on Canadian dinosaurs, Parks summed up the opportunities available for the future.

> The collecting, mounting, and preserving of these great relics of the past is a work involving much labour, time, and money. It is however, a work of national importance and one that cannot be deferred to some future time. The ceaseless erosion of the fossil-bearing beds, here and

> there and from time to time, exposes a skeleton; if this skeleton is not collected it succumbs to the wear and tear and is lost forever. The Dominion, the provinces, and the universities should unite in an effort to organize annual expeditions to prevent so great a loss.

Parks' words might have been visionary, but they were not (at least in the short term) prophetic. In the year in which they were published, Mussolini invaded Abyssinia, and Hitler reintroduced compulsory military service. At home, Canada and the U.S. were in the middle of the Great Depression, which caused the National Museum fieldwork to cease for several years from 1930. In the U.S., some workers on relief were employed to do important dinosaur work, but it probably never crossed the mind of anyone in Canada that the "On to Ottawa" marchers could have been paid to dig dinosaurs on the Red Deer River. The Second World War sidetracked younger palaeontologists, and on their return some moved out of the field altogether (like Ray Sternberg) or turned their attention to other kinds of fossils. With hindsight, Parks' words better describe the decades before he wrote than the ones that immediately followed.

Palaeontology was itself changing. Major new dinosaur discoveries were being made in the world beyond North America. Dinosaurs were being heavily hyped in American museums and their periodicals, and as they grew in public appeal, they were losing their appeal to the young Turks of palaeontology. Dinosaurs themselves were seen as sluggish, rather static reptiles, about which a lot was known, and the beginning palaeontologist perhaps felt it would be easier to gain a reputation by studying something else.

But in different parts of Canada, interest was growing. In the west, communities were increasingly aware of the dinosaurs from their region, and were trying to feature them in museums, parks, and zoos. Back east, the bigger museums were finally turning their attention to popular literature about dinosaurs, and the first film on dinosaurs had appeared. It is time to look at the dinosaurs as a public presence, before returning to the scientists' story after the Second World War. And the American Museum was, once again, the leader.

5

"WE HAVE TO HAVE SUPPORT FROM THE PUBLIC"

"One cannot be a vertebrate paleontologist . . . nowadays in an ivory tower. We have to have support from the public at large and from the government in order to do that. . . . The museum I come from has very large displays of fossil vertebrates . . . and, of course, it is something of a chore to have to bother about these displays. Nonetheless, it is interest in our displays, in our dinosaur halls, that has brought support, very tangible support, into the museums and has allowed us to go ahead on other problems that are of interest to various members of the institution."

EDWIN COLBERT

OUT OF THE IVORY TOWER

"My kid loves dinosaurs," says almost everyone I meet. But as late as the early 1950s (when I was in high school in England), most children and adults didn't really know what dinosaurs were, and if you were interested (as I was), information was hard to come by. Few books or movies featured dinosaurs; if you had read *The Lost World* it was the book by Sir Arthur Conan Doyle (better known for Sherlock Holmes), not Michael Crichton. Parks featuring dinosaur sites either did not exist or were mainly in the western regions of the North American continent, far from centres of population and (before good roads and cheap gas) inaccessible to most people. Occasional newspaper stories discussed major finds, but dinosaurs received little publicity.

But by mid-century, many dinosaur fossils had been brought into museums — even in Canada. Laboriously prepared specimens were on display, and public interest was rising. Curators had published many papers describing their discoveries, but these were seen only by other specialists. Curators typically felt most comfortable in an ivory tower, writing papers for other specialists and unconcerned about the public at large. If they tried to publicize their research, they sometimes had to convince senior administrators that they were not wasting their time.

A few museums realized that they had a potential gold mine, and the work of enlightened curators, educators, and publicists gradually began to build the public image of dinosaurs that is familiar today. And out where dinosaur bones were found, some public leaders became fascinated by dinosaurs, gathered information about them, and used their newfound knowledge. Some became amateur collectors; some became promoters who tried to feature dinosaurs in their community. And, gradually, they began to influence politicians — the controllers of the public purse.

It is time to look back and see how a few leading museums nourished the beginnings of public interest in dinosaurs in Canada, and how in turn that interest affected regions in which dinosaurs were found.

Mural on the Drumheller Fossil Museum.

"INTEREST IN OUR DISPLAYS"

The American Museum of Natural History took an early lead in both dinosaur research and (when the scientists could be dragged away from new finds) exhibition. From a public perspective, display is what dinosaurs are all about. When we enjoy dinosaurs in books or movies, many of us mentally link to the "real thing" seen in museums, where we have seen huge skeletons, marvelled at the diversity of form, and boggled at bizarre shapes.

The sheer size of many dinosaurs presents a space problem. Whether the need is to store tonnes of bones in plaster jackets, lay out all the bits of one skeleton, or display it as an open mount, museum planners — and fund-raisers — have had to rise to the challenge. Indeed, dinosaurs have had a major impact on museum design, requiring wide corridors, large elevators, big labs and storage areas, and even bigger exhibit galleries.

Technicians were pressured to get specimens out of the labs as soon as they were completed, so each new skeleton went on display as soon as it was ready, while others were added piecemeal until a complete gallery could be completed. Exhibits of Alberta dinosaurs were opened in New York (1913 onwards), Ottawa (1913 onwards), Toronto (1920 onwards), and London (1926).

Reasonably complete dinosaurs with strong bones were made into open mounts, whose skeletons were supported in what was regarded as a natural position. Technicians learned to create armatures of welded iron rods cradling or even penetrating the bones. A simpler style, variously called a plaque, panel, or slab mount, was used for more fragile dinosaurs, or ones in which only one side was presentable. These often showed the bones in the way they were found. Actual dinosaurs were interpreted by labels (usually terse and technical), sometimes supplemented by maps, diagrams, and perhaps paintings of the animal's imagined appearance in life.

"A HUNDRED MARVELS YOU MAY SEE"

Matthew celebrated Osborn's American Museum dinosaur exhibit in rhyme.

> . . . But now old Barnum is outclassed,
> A greater showman's come at last.
> In Osborn's great menagerie,
> A hundred marvels you may see.

Albertosaurus *has a leg to stand on.*

At the American Museum, the finest specimens, such as *Albertosaurus* and *Corythosaurus*, were displayed as soon as possible. The duckbill *Saurolophus* became a panel mount in the gallery by 1913. By 1915, technician Peter Kaisen had mounted the little carnivore *Struthiomimus* as a panel mount, in the position in which it was found, with its head and neck bent backwards over the body as the muscles dried after death.

By 1917, the American Museum had two complete dinosaur halls. One focused on Jurassic dinosaurs from the American west, and the second was the world's best exhibit of Cretaceous dinosaurs, using material from Montana and Alberta. Photos of the Cretaceous gallery show skeletons in rows, as if frozen on the march. *Tyrannosaurus rex* is knocking his head against the ceiling, among the light fixtures. Glass cases between the skeletons show smaller items, and occasional paintings on the walls present the palaeontologists' and artists' vision of living dinosaurs in their environment. Many of these dinosaurs are still on display; the museum's Great Hall of Dinosaurs, which was finished in 1995, is described in a recent book in which more than one-third of the featured dinosaurs are still from Brown's expeditions to Alberta.

Other American museums acquired Canadian dinosaurs, either by purchase (like Sternberg's material in San Diego), or by exchange between museums. Specimens (mainly collected by the American Museum and the Royal Ontario Museum) went to the Field Museum in Chicago, the Smithsonian, Los Angeles, and the University of Michigan.

Other Canadian dinosaurs were displayed in South America — in Buenos Aires, La Plata, and Rio de Janeiro. Meanwhile, the British Museum's *Scolosaurus* was mounted and displayed in 1926, in such a way that the upper and under sides can both be seen, showing off its remarkably complete skin. For many years it was the only Canadian dinosaur on display in Britain.

"IF OTTAWA DOES NOT WANT THE DINOSAUR"

The Geological Survey had the *Albertosaurus* skulls found by Tyrrell as well as Weston's skulls "on exhibition in its museum for some years" by 1903. At that time Lambe directed attention to "the desirability of mounting in a permanent and attractive manner those specimens of the vertebrate collections that have been recently described and figured, and of providing space for their exhibition to the public."

Completion of the Victoria Memorial Museum building in Ottawa in 1911 at last provided adequate exhibition space. The Survey started moving into the building in 1910, but when the Sternbergs arrived in 1912, they "found there a great museum building with little original material for exhibition, chiefly casts." But the Sternbergs brought in "a priceless collection of . . . monsters," and the Hall of Fossil Vertebrates was one of the first to be opened, on January 20, 1913. It was immediately popular, being especially crowded on Sunday afternoons. In 1913 there was a report that the duckbill eventually called *Edmontosaurus* was "now being mounted in high relief preparatory to being placed on exhibition in the museum of the Geological Survey, Ottawa." The fossil became "noteworthy as the first dinosaurian specimen to be mounted for exhibition in a Canadian museum."

Only three years later the Houses of Parliament were destroyed by fire, and the new museum was chosen for temporary quarters. The fossil exhibition hall was cleared at a few days' notice: "we shoved everything up against the wall, covered them up," remembered Charlie. This, it turned out, was to make room for the Senate — a joke that tempered the frustration of the museum staff as they were moved to other buildings, or into the basement. Palaeontology was the only scientific division that remained in the museum. "One day someone in a high position had his office directly over my lab and one day he found a louse and he sent his assistant downstairs to ask me if any of these dinosaurs had lice. Well, I said, 'you can tell the honourable gentleman I haven't anything here that is less than 3 million years old. Let him draw his own conclusions.'"

When Parliament left the museum in May 1920, the palaeontologists felt that insult was added to injury. The fledgling National Gallery was given extra space so that it occupied the entire East Wing, including the space formerly occupied by vertebrate palaeontology. Charlie records that the director of the museum learned of this plan only by chance: "You can see why we had a little trouble getting things going."

The horned dinosaur skeleton was considered to be too fragile to be moved easily, and there was a suggestion that it should be boxed up and left among the paintings. The *Canadian Mining Journal* was scandalized, and suggested: "If the right of the Palaeontological Branch to continue its exhibits in the Victoria Museum be taken away, then it were a pity to have disturbed the dinosaur from its age-long rest in Alberta. . . . Perhaps, if Ottawa does not want the dinosaur, the province of Alberta, or the United States, might welcome it. That is where most of the geologists have gone anyway." During the hasty evacuation specimens were scattered and mislaid, or became separated from their labels.

Through this period of disruption, other specimens (such as the larger skull of *Cheneosaurus* described in 1917) were prepared for exhibition as they became available, but it is not clear whether space was found to show them. Later, in 1926, skeletons of two horned dinosaurs (*Chasmosaurus*) were exhibited side by side in the museum: Ottawa's first attempt at a group mount of horned dinosaurs. Although formally designated as Canada's National Museum, the institution remained firmly under the care of the Geological Survey until 1947, when it finally achieved a more than nominal independence and enhanced funds.

At the Royal Ontario Museum, dinosaurs were first exhibited when at "an impressive ceremony" in 1920 a panel mount of the duckbill *Kritosaurus* "was introduced to the citizens of Toronto." Other dinosaurs were put on display as they were ready; for instance, Parks notes in 1922 that excavation and mounting of *Parasaurolophus* in the Royal Ontario Museum of Palaeontology was done "under the supervision of Mr Levi Sternberg." In 1926 he recorded that *Prosaurolophus* "has been skilfully mounted by G.E. Lindblad and is now on exhibition."

A new wing was added, and the museum reopened in October 1933. Vertebrate palaeontology now had four galleries and offices, and associated workrooms. By this time the dinosaur gallery had ten complete skeletons, and between fifteen and twenty partial skeletons and skulls, about half of which represented new genera or new species of recorded genera. At the time this display was regarded as "second only to New York," which at that time meant second best in the world. Photographs from this period show panel mounts, free-standing skeletons of *Chasmosaurus* and *Corythosaurus*, and a painting by G.A. Reid showing a "generalized landscape of Upper Cretaceous Time."

EXTREMELY ENTHUSIASTIC GROUPS

At a conference of palaeontologists held in Edmonton in 1965, British fish specialist T.S. Westoll spoke enthusiastically of the public relations of palaeontology. He was "immensely impressed by what is possible . . . within the province." He praised a school collection at Eastend, Saskatchewan, and the local museum at Drumheller, that "show the results of local collecting by extremely enthusiastic groups and I think that the public impact of this sort of thing is very fine and is possibly unequalled." During the decades following the dinosaur rush, the fossil collectors of the west — in best pioneer tradition — had "pulled themselves up by their own bootstraps" to the extent that visiting specialists could be impressed.

Nevertheless, if you wanted to see a western Canadian dinosaur at the end of the 1930s there were any number of places you could go — London, England, for instance; Buenos Aires and other South American capitals; New York, Washington, Chicago, and California; and in eastern Canada, Ottawa and Toronto. Of course, if you lived in Alberta or Saskatchewan, you would be out of luck.

Even after that, for some forty years the best-known visual images of dinosaurs in Alberta were two life-size (if not very lifelike) sculptures — one in Calgary and one in Drumheller. With the western boosters' reverence for size, both were BIG. It seems to have been irrelevant that both were of dinosaurs that (at least at the time they were created) had never been recorded in Canada.

These collections and sculptures at first were essentially amateur efforts, but this designation is not intended to be a criticism. In fact, Canada's dinosaur palaeontology owes a lot to dedicated amateurs; farmers, tradesmen, teachers, and mechanics live in the country where the dinosaurs are found (or drive many kilometres for weekend visits).

The least of them are sharp-eyed rockhounds able to pick out an unusual bone. Many are knowledgeable, enthusiastic, dedicated amateur fossil collectors; some of them could have found a career in palaeontology had their circumstances been different. They collect dinosaur bones and send them to specialists, or even become capable specialists themselves. They lead the professionals to their finds, and often provide meals and company in the lonely hours, days, and weeks in the field. They are also the source of enthusiasm and information driving local museums and other educational efforts.

CORKY, CHARLIE, AND OLD HORNYFASE

A representative amateur is Harold "Corky" Jones (1880?–1978). Son of Queen Victoria's doctor, Corky was born in England, and his father took him to fossil beds on the Isle of Wight while he was still a child. He had only limited schooling, but "having bones around, he could not be content until he got books and learned what they might be." At eighteen, Corky wanted to be a cowboy, and moved to Eastend, Saskatchewan, in 1898. He collected fossils where he could find them, and information when he could get it.

In 1921, Charlie Sternberg saw a horn core of *Triceratops* in an office in Eastend, traced it to Corky, and commented, "this was a most important discovery because it proved the presence of beds of Lance age in an area where they were not previously known." The two began a correspondence in which (as well as dinosaurs) Corky sent fossil plants, mammals, and artifacts to the museum.

In February 1930 Corky was inquiring if Sternberg had "the Triceratops specimen mounted yet," and asking for a photo. Charlie replied that "old hornyfase [sic] is not yet unpacked." In 1931 Charlie returned a *T. rex* bone sent for identification, and commiserated that "things are in such bad shape out there."

In the 1930s, Loris Russell met Corky. At the time, "he was town constable and had installed a wooden post in the main intersection to dissuade vehicles from cutting corners. This became known as Corky's police force, and when on Sunday morning it was found leaning at an inebriate angle, this dereliction of duty would be gravely reported to the 'Chief.'" In 1932 Corky predicted that the west "was going to have another disastrous year." He reported to Sternberg a visit from Russell, in which he "showed me where I had no less than three different animals represented in about five feet of dinosaur tail."

In 1936 Corky reported another find: "... all that showed was a ball-joint.... we were surprised to find nearly a whole *Triceratops* skull stuck on the end of it." Sternberg set out the basic collecting techniques in a two-page letter, and Corky not only learned how to collect, but also how to prepare and identify his finds.

Things slowed down during the Second World War. "I have done little fossil hunting the past few years, having no car of my own and the gas and tire shortage I have had no opportunity. I have also been told by the doctor to go easy."

Sternberg was travelling west in 1946, and wrote to Corky asking for him to locate tent poles and supplies of plaster. Corky replied that both were available. "He has 6 tons of Hard Wall plaster on hand and I can only hope your luck while here will be so phenomenal that you will use it all up." In 1947 Corky collected a new ceratopsian crest. "It is definitely not Triceratops and I have no picture of anything quite like it.... After I had it mounted I could not buy a film and had to wait until a friend went to Regina to secure one for me." Eventually he was able to send photographs. Three years after Corky's death the fossil was finally published as *Torosaurus,* a first for Canada.

These and other fossils are on display in the museum at Eastend, which for a long time was in the old school basement. "We started this collection," Corky reminisced, "as so many specimens were being taken out of the country and given to the large museums.... these creatures, lived, died, and got buried and resurrected here, and this is the logical place for them to be kept, and viewed by those wishing to see them." In 1952 the school was flooded, but many of the specimens were rescued.

Tributes to Corky include many comments from colleagues. "In effect he was an educator without salary throughout a great part of his life," wrote Bruce McCorquodale of the Saskatchewan Museum in Regina. Wallace Stegner, the well-known writer who grew up in Eastend, also celebrated his work. "If a community is ... like a pile of kindling, inert and heatless until some accident of heat or some great man touches it with fire, then Corky, in his humble and unpretentious way, is a sort of light bringer. Corky's example can teach [any Eastend child] that it is possible to be a contributor to civilization even from a prairie town."

CALGARY'S PREHISTORIC PARK

Calgary had a museum as early as 1911, "located in a decrepit downtown office-warehouse district adjacent to skid row and one block removed from the tacitly-accepted brothel district." Dinosaur bones

Dinny stands proudly at the Calgary Zoo.

were included in its main floor display, and presumably Cutler's early collections were added to these. The museum closed in 1937 as an economy measure. (When Calgary eventually opened a new museum, the Glenbow-Alberta Institute, its scope was limited to art and human history, so palaeontology was not represented in its exhibits.)

The Calgary Zoo on St. George's Island eventually got under way in 1927. The zoo's society took an interest in the finds along the Red Deer River, and often discussed them at meetings. In 1931 Director Lars Willumsen visited Hagenbeck's Zoo in Stellingen, near Hamburg, Germany. Animal collector Carl Hagenbeck was an imaginative entrepreneur, a sort of German Barnum. One of his expeditions to Africa had been made in search of living dinosaurs reputed to have been seen there; perhaps as some compensation for his lack of success he had created a series of models of prehistoric beasts for his zoo. Willumsen was thrilled, and returned with photos, models, and ideas. By 1932 a committee had been established to construct a Prehistoric Park in the Calgary Zoo. The zoo's founder, colliery-owner Dr. O.H. Patrick, started a palaeontological park in 1932, importing carloads of dinosaur bones and petrified wood and building two "fossil houses" in which to display the fossils. When the Calgary museum closed, its dinosaurs ended up in the zoo.

Expertise was solicited from Charlie Sternberg, Barnum Brown, and (by correspondence) William Swinton of the Natural History

Museum in London. Sculptor Charles Beil from Banff worked with Arne Koskeleinen and John Kanerva at the zoo.

During the 1930s, a series of fifty models of dinosaurs and other prehistoric beasts was constructed out of cement. The first was Alberta's *Chasmosaurus*, but the series represented important fossils from many areas, including the American Jurassic *Stegosaurus*. Last to be built was a spectacular "Brontosaurus" (now *Apatosaurus*), 35 feet (11 metres) tall and weighing 120 tonnes. Nicknamed "Dinny," this monster was unveiled in 1938, and became the zoo's mascot. (Although many years of research have reduced our expectations of finding sauropods in Canada, the possibility was still considered as late as the 1940s, when an unsigned national museum article suggested that "It is likely that these strange creatures also roamed the badlands of Alberta. . . .")

In 1956, the Alberta Society of Petroleum Geologists offered to help move the models into chronological sequence, so that the visitors could trek through time. Forklifts and a Royal Canadian Air Force crane rearranged the models, and they were repaired and repainted. The following year Dr. Swinton assured the zoo that it had the most up-to-date dinosaur display in the world.

DINOSAUR TRAIL

Drumheller has drawn its sustenance from the rocks, for it began life as a coal-mining village in 1913, and at its peak in 1948 there were over thirty active mines. Even while the mines flourished, a dedicated citizens' group became concerned about dinosaurs. Rockhounding is a strong local tradition, and local collectors banded together to put on a public display in a room adjacent to the Rotary Swimming Pool during the 1940s and 1950s. W.R. Reade, Jessie Robertson, and others formed the Drumheller and District Museum Society in 1957, and its museum opened in 1960, one of the few community museums in North America to focus on palaeontology.

A radical expansion followed the loan of a nine-metre (thirty-foot) skeleton of the duckbill *Edmontosaurus* from the National Museum. During my early visits to Drumheller in the 1970s, the vivacious Jessie Robertson (in her eighties) was still much involved with the museum; it was easy to tell if she was on duty, as her splendid antique auto was often parked outside. She would talk enthusiastically to visitors of what she called "my dinosaurs."

Local boosterism went beyond the museum as the city developed a reputation as Alberta's "dinosaur capital." In the 1950s a circular route was planned to take tourists through the badlands north and west of the city. Called the Dinosaur Trail by the citizens, and a "lesson in human survival" by writer Robert Kroetsch, it led along the north side of the Red Deer River, climbing the bluffs and zigzagging through the fields. It swooped down to the river at Bleriot Ferry (named for a brother of pioneer aviator Louis Bleriot, who managed it for many years), climbed out again, and ran back along the south side. The roads were gravel, and often in poor shape — I remember camping with a group of naturalists at the ferry in the 1970s and being unable to get up the hills on either side after heavy rain until the road dried out.

Perhaps under the influence of the Calgary Zoo, local Tig Seland (later president of the museum society) constructed a life-size (but otherwise eccentric) *Tyrannosaurus* model out of concrete and chicken wire. Gradually others added their own models, and today the town, the trail, and the rodeo grounds are liberally decorated with a variety of model dinosaurs, of various kinds and qualities. An annual publication, *The Badlands of the Red Deer River*, for years promoted the dinosaurs and included a wide variety of entertaining, and sometimes accurate, information.

Edmontosaurus *at Drumheller's fossil museum.*

TREASURE-HOUSE OF THE AGES

In 1915, within a few years of Brown's appearance on the Red Deer River, Dr. Winfred George Anderson (1881–1966) began agitating for preservation of the area that is now Dinosaur Provincial Park. Anderson was born in Stratford, Ontario, graduated from the University of Western Ontario, and opened a practice in Granum in southern Alberta in 1904. He came to a homestead at Wardlow (east of Steveville and north of Brooks) in 1911, and lived there until the 1950s. He visited the Sternbergs' field camps around 1913, and raised the idea of a park with the local newspaper as early as 1914. Initially he lobbied Ottawa, but when in 1930 responsibility for natural resources was transferred to the prairie provinces, he shifted his attention to the Alberta government. In the same year the Alberta government passed a *Provincial Parks and Protected Areas Act*, which included provisions for protection "of objects of geological . . . interest," and began to establish parks in 1932. But money was tight in the Depression, and parks were low priority.

In 1935 Charlie Sternberg began to map the sites (many from memory) from which dinosaurs had been taken in the Steveville area. He continued with the assistance of his son Raymond in 1936, and the quarries were marked with steel pins with bronze caps set in concrete. In 1938 the first formal action was taken to protect dinosaur sites. The Alberta government passed an Order in Council protecting areas around the Munson and Morrin ferry sites, but it seems never to have been enforced.

In 1950 the Geological Survey published its geological map of the Steveville area, featuring the astonishing number of sites that Charlie had identified. Anderson organized a 1954 public pilgrimage to Sand Hill Creek, where Levi Sternberg was excavating in the rain. A *Lethbridge Herald* editorial urged that "a park . . . would provide machinery for preserving this treasure-house of the ages . . . and it would provide facilities for the people to see, study and enjoy this caprice of nature. . . . So far the province has done nothing except to 'pass the buck' to Ottawa."

Finally, in 1955 (sixty-six years after its importance had been recognized), the Steveville-Dinosaur Provincial Park was established. "The government approached me to come out to look the situation over and make a recommendation," remembered Charlie Sternberg. "'From the Red Deer to the border,' I said." Initially the park's size was 8936 hectares (22,072 acres), but some of the land designated was private rather than Crown land, and later these boundaries were narrowed. The park did not open to the public until 1959.

The park's first warden was a fortunate choice. Roy Fowler (d. 1975) was a farmer from the High River area south of Calgary, who was a capable amateur naturalist and geologist. Loris Russell "first encountered him in 1928. . . . [He] had a shed full of fossils and was very generous with them. Later, when Roy had the opportunity to move to the park, I strongly recommended him as the first custodian. He was enthusiastic, without a formal training, but was self taught."

Fowler was concerned that a dinosaur park should show dinosaurs, and found a skeleton in the summer of 1955. The recently retired Charlie Sternberg found another, and these and another later one were preserved *in situ,* protected by small fossil houses. By 1962 Steveville — the little town remembered by all the dinosaur hunters of the dinosaur rush as a source of supplies, mail, and hot baths — had become a ghost town. In that year it was dropped from the name of the park, so that people would not look for the park near a non-existent town.

There was a move afoot to name the park after Anderson, but he refused the honour. He retired to Medicine Hat, and when he died in 1966, his ashes were scattered in the park.

POPULAR PUBLICATIONS

Today, when every bookstore has a shelf of dinosaur books, it is hard to envisage a time when there were no popular books of any kind on dinosaurs. Dinosaurs were of course mentioned in the few non-technical books on fossils, and described in museum leaflets; Matthew's 1915 guide to the American Museum's dinosaur collections seems to have been the first book to focus on dinosaurs exclusively. More commercial publication began with William Swinton's *The Dinosaurs,* published in the U.K. in 1934, and Edwin Colbert's *The Dinosaur Book* in 1945. As the flood of discoveries arrived at a few museums, the many people who could not visit these collections relied on newspapers and magazines for information. Newspapers (which in those days did not have science correspondents) were notoriously inaccurate, so magazine articles, popular handbooks, and books written by scientists themselves were of particular importance.

Brown had only completed his first full season on the Red Deer River when he put pen to paper to write a popular account. "Fossil Hunting by Boat in Canada" appeared in the *American Museum Magazine* (an ancestor of the present-day *Natural History*) the following year. As fast as Brown's scientific descriptions went to press and the specimens went on display, a short popular article saluted the new find. Brown's account of *Saurolophus* was featured as "A New Crested Dinosaur," with half a dozen photos showing the raft, the skeleton, the panel mount, and the work of excavation. However, the flood of material kept Brown busy with scientific descriptions, and he could not keep up the popular articles.

When he came to Canada, Charles Sternberg had already published *Life of a Fossil Hunter* (1909). In 1917, his second book appeared; *Hunting Dinosaurs in the Bad Lands of the Red Deer River, Alberta, Canada* was the first book to focus substantially on Canadian dinosaurs. Sternberg published five hundred copies at his own expense, sold some, and used others for promotion of his fossil-collecting activities. This second book picks up Sternberg's story from the end of *Life of a Fossil Hunter,* and includes eight chapters on his Alberta work. In a rambling but entertaining fashion, Sternberg talks about his family's work in Alberta, gives flashbacks to early discoveries, and throws in some verse and fiction. Both books went through further editions in Sternberg's lifetime, and have been reprinted since.

Brown had given up on describing his discoveries piecemeal. In 1919, a lengthy article, "Hunting Big Game of Other Days," summarizing key events of his Alberta expeditions, appeared in *National Geographic.* Brown used the analogy of big game hunting, and presented a picture of fossil hunting that could be grasped by relatively unsophisticated readers.

Brown describes the country and outlines the geology, and then calls up a picture of the landscape occupied by the dinosaurs — "a vast stretch of jungle-covered delta and coastal swamp, interspersed with bayous and lagoons" — comparing it to the Everglades of Florida. Having carefully set the scene, Brown then discusses all the dinosaurs together, making few distinctions of age. He talks of the duckbills *Trachodon* and *Saurolophus,* the bird mimic *Ornithomimus,* horned dinosaurs *Monoclonius* and *Triceratops*, and the armoured *Ankylosaurus.* These are preyed on by the flesh eater *Albertosaurus.*

After a brief acknowledgement to the Geological Survey's work in determining the geological formations and securing "a variety of fossils," Brown then describes his methods of working during the "many years" the American Museum "has been making a systematic collection of fossils along this river." However, he refers only to the results of four seasons, so the article seems to have actually been written in 1913 or early 1914; perhaps its completion was sidetracked by war activities. He ends with a dramatic reference to the future promise of the area: ". . . the field has by no means been exhausted . . . , and for all time to come the Red Deer River will be a classic locality for collecting prehistoric treasures."

The article is illustrated by photographs and reproductions of paintings. The photographs show river valley landscapes, geological sections, and excavations. Other pictures show the difficulty of transporting plastered bones — a couple of horses pulling a sled with a 270-kilogram (600-pound) block containing the hip bones of *Hypacrosaurus,* and another block being lowered down a steep slope, guided by a fossil hunter standing directly under the block (who is no doubt hoping that the rope would not break). The most evocative and often-reproduced picture shows a six-horse team pulling a loaded wagon up a steep "cow trail" that is recognizable as the main road that now leads into Dinosaur Provincial Park. The paintings are restorations of the living animals by the great palaeontological artist Charles Knight, and feature both Alberta dinosaurs and a job lot of extinct animals not mentioned in the article.

Brown's Calgary contact, William Pearce, kept an eye on dinosaur-collecting activities. In October 1922 he wrote to Rankin, head of the Publicity Branch of the Canadian Pacific Railway, that "A gentleman connected with McLeans [sic] Magazine . . . had been out north-west of Brooks where the dinosaur skeletons are found. . . . he seemed surprised that the CPR had not drawn public attention to this field, and claimed that it would be a veritable drawing card . . ."

The CPR was in the tourist business, and had already built the Rocky Mountain resorts of Banff and Lake Louise into major international resorts, so they were not likely to neglect another opportunity. Rankin agreed that "regular publication of an article is an excellent one, and I will act on it shortly." Pearce suggested that Barnum Brown be approached, leading to a booklet entitled *The Dinosaurian Remains in the Red Deer River Valley of Alberta.*

WHAT STRANGE ANIMALS LIVED IN ANCIENT TIMES

Canadian museums also tried to bring their dinosaurs to a wider audience. In 1921 Charlie Sternberg published "A Popular Description of Dinosaurs" in the country's leading nature magazine, *Canadian Field-Naturalist.* In 1929 an account of "Dinosaurs in the National Museum of Canada" was shared with his fellow civil servants in the *Civil Service Review,* and a popular account of his research on "Prehistoric Footprints in Peace River" was published in the *Canadian Geographical Journal* in

1933. In 1945 Sternberg went to the same periodical to publish "Canadian Dinosaurs, their place in nature and the conditions under which they lived," and in the following year *Canadian Dinosaurs* appeared as a National Museum Bulletin.

It is an indication of how unfamiliar this material was to the general public of that time that Sternberg found it necessary to explain how to pronounce the word "dinosaur." He takes twenty pages to explain what is known of dinosaurs (particularly Canadian ones). Popular errors are disposed of, and Sternberg points out that the "strange animals of ancient times" are so because we are unfamiliar with them, while bats, giraffes, and turtles would seem bizarre if they were only found as fossils. A palaeogeographic map shows Matthew's vision of North America in Late Cretaceous time, with an inland sea running from the Arctic to the Gulf of Mexico, and swamps on its western margin in which the dinosaurs had lived. Sternberg takes us, as if "by some mysterious time machine," to the ancient landscape. Seventeen illustrations include diagrams, and photographs of landscapes, excavations, dinosaurs in the field, mounted dinosaurs in the museum, and models and paintings prepared for the museum. Diagrams are borrowed from the Royal Ontario Museum and the American Museum of Natural History, but the photographs are from the splendid series taken by survey geologists since Dawson's day.

At the Royal Ontario Museum, Parks also produced popular articles. In 1919 he described *The Great Fossil Reptiles of Alberta* for the Hamilton Association, and in 1924 published (in forty-four pages) *Dinosaurs of Alberta* for the University of Toronto Press. More than a decade later, in 1935, he described "Dinosaurs in the Royal Ontario Museum" for the *University of Toronto Quarterly.*

CHARLIE — A DOMINION MOTION PICTURE

As early as 1912, *Gertie the Dinosaur* was the subject of a cartoon, and the documentary makers were not far behind. In 1921, the Dominion Motion Picture Bureau began a film of the Geological Survey's dinosaur activities, filming Charlie in the lab in the National Museum. The field crew had obviously been told to film Sternberg, but did not expect to find two independent Sternberg parties working in the same area. The locals directed the film crew to George, working for the University of Alberta, and the film shows him using explosives to remove overburden, the crew scattering as the charge goes off. The lab sections of the film were made of Charlie in the National Museum.

Parts of the footage were reused in a later National Film Board movie about the life of Charlie Sternberg, simply named *Charlie*. Such isolated instances foreshadow the role that dinosaurs would come to play in modern educational television.

By mid-century, most professional vertebrate palaeontologists were losing interest in dinosaurs, turning their attention to neglected groups. Ironically, as their museum galleries filled up with dinosaurs, public interest was growing. And in 1947, another underground resource was to play an important role in the development of western dinosaurs. In 1947, at Leduc, south of Edmonton, an oil well went wild. It was the beginning of a huge oil boom in the west, which was to make funds available, eventually, for dinosaurs.

6

DINOSAUR RENAISSANCE

"Most contemporary paleontologists have had little interest in dinosaurs; the creatures were an evolutionary novelty, to be sure, but they did not appear to merit much serious study because they did not seem to go anywhere: no modern vertebrate groups were descended from them."

ROBERT BAKKER

"A BIT PASSÉ?"

In the 1970s, while interviewing for the Provincial Museum of Alberta, I asked each of the candidates how they felt about dinosaurs. One, with an air of world-weariness, suggested that they were perhaps "a bit passé." Although this view was not shared by other candidates, it was not uncommon at the time among vertebrate palaeontologists. Dinosaurs had been extensively studied for more than a century, and seemed to be fairly well understood.

For most of the century since Dawson's first western dinosaur find, Alberta had neither a vertebrate palaeontologist nor a provincial museum. The province had shown little interest in its heritage beyond ranchers, cowboys, and Mounties, and tended to leave such activities to local enterprise. Other than the valiant efforts of John Allan, the university had done little to fill the gap. Yet British Columbia established its provincial museum in 1887, and even impoverished Saskatchewan had a museum of natural history by 1950.

Leduc #1 and succeeding oil discoveries began to change Alberta from a "have-not" to a "have" province; by the 1960s change was in the air, and money was available to fund areas of public life that had hitherto been neglected. There was a willingness to look again at minority interests, and leaders were ready to seize new opportunities.

TO LAUNCH VERTEBRATE PALAEONTOLOGY

For a few days from August 29 to September 3, 1963, Alberta hosted a remarkable gathering of people interested in vertebrate fossils. Financial support from the Alberta government, the University of Alberta, and other bodies made it possible to bring a group of eminent specialists from the International Congress of Zoology to meet with key people in Alberta. The conference was instigated by the university, where "some of us. . . . had felt for some time that an active program in Vertebrate Paleontology should be included amongst the University's activities. This was the *raison d'être* of our conference — to bring experts to Alberta who could advise us on what should and could be done to launch Vertebrate Paleontology as one of the major activities of our University."

Among the thirty-four attendees were international experts of the stature of Edwin Colbert (dinosaur expert at the American Musum of Natural History), Alfred Romer of Harvard (whose text on fossil vertebrates was the standard work for several decades), mammal specialist Bjorn Kurten from Finland, and fish expert T.S. Westoll from the U.K.

The National Museum was represented by Charlie Sternberg (retired from Ottawa but still active) and his successor Wann Langston, Jr. (who had already returned to Texas). C.S. (Rufus) Churcher, Gordon Edmund, Loris Russell, and William Swinton were present from the Royal Ontario Museum.

Among those from Alberta were Robert Folinsbee, Charles Stelck, and Donald Taylor of the University Geology Department; Lu Bayrock from the Research Council; and Potter Chamney of the provincial Geological Survey. The new provincial-museum-in-planning was represented by Ray Harrison (who became its first director) and Bruce McCorquodale (then of the Saskchewan Museum of Natural History, but another future director of the Provincial Museum). The citizens' community was represented by talented painter and palaeontologist Hope Johnson of Ralston (near Medicine Hat) and Dr. Read and Mrs. Westergard of the Drumheller Museum.

The conference was just over a decade short of the centenary of the discovery of dinosaur remains in Alberta, and Wann Langston's summary gave some impressive statistics. Nearly two hundred technical papers had been published on Alberta vertebrate fossils, including records of forty-five genera and seventy-nine species of dinosaurs — forty percent of the known Upper Cretaceous dinosaur faunas of the world.

There were hints of new directions — the technique of washing and sieving large quantities of sediment had shown in Wyoming that dinosaurs were only five percent of the total fauna, and "Enough small material has been picked up from the surface of Cretaceous outcrops in Alberta to suggest that similar procedures could be equally rewarding here." The group bussed around key sites in the province, found some mammal fossils at Scollard, and had a steak barbecue with the Brooks Chamber of Commerce in Dinosaur Park.

In the general discussions that followed in Calgary, Dr. Westoll encouraged the development of the proposed provincial museum, and urged a modest program in the university to train specialists in the field. "Everybody nowadays is interested in dinosaurs," suggested Colbert, who felt that a university program would be in a position to carry out work on "some of the most spectacular dinosaur faunas in the world." His views of the provincial museum were modest: "A series of dinosaurs, two or three. . . . in the nice attractive exhibition hall in the provincial museum, would create a tremendous amount of interest in Vertebrate Paleontology in the province and in neighboring provinces."

By the time the conference proceedings appeared in 1965, the planned appointment of a vertebrate palaeontologist to the university staff was announced, and it was hoped that the "Provincial Museum will eventually be able to display specimens and add greatly to the public interest in Vertebrate Paleontology."

Although dinosaurs were a large part of the excuse for the conference, the discoveries (and the papers published with the proceedings) were of mammals, and scientific enthusiasm was for the areas neglected during a century of dinosaur studies. The conference certainly fulfilled its objective of providing academic support for expanded programs in vertebrate palaeontology, and the institutions and programs would be established on a larger scale than anyone there had envisaged.

No one present at the conference is likely to have predicted that within a decade dinosaurs would again be one of the hottest topics in vertebrate palaeontology. In their wildest dreams no one there would have imagined that within fifteen years the Alberta government would begin work on the world's most spectacular fossil museum.

THE DIM AND DISTANT PAST

In 1919, Barnum Brown put what he called the "dim and distant past" of the Cretaceous no further back than three million years. We now know the period ended some 65 million years ago.

The drama of dinosaur discovery has been played out against another search, for a means of measuring the age of the earth. Until the beginning of the nineteenth century the standard was Archbishop Ussher's date for creation of 4004 B.C., calculated by adding up generations of "begats" in the Bible. More scientific methods were developed through estimates of the time needed to deposit great thicknesses of strata, and for a presumed molten earth to cool to its present state. Not until the beginning of this century did scientists realize that heat generation through decay of radioactive minerals affected these calculations, and that the proportions of radioactive minerals to their decay products provided a more objective means of measuring the actual age of rocks.

Lengthy practical and theoretical discussions ensued before current views developed in the 1960s. Since then, Earth has been seen as approximately 4.5 billion years old, and dates for the Mesozoic era have varied only a little. The Triassic period is now considered to have begun 245 m.y. ago, the Jurassic period began 208 m.y., and the Cretaceous period lasted from 145 to 65 m.y.

THE REMARKABLE LORIS RUSSELL

The leading figure in the Royal Ontario Museum's contingent was Loris Russell. His career began in the 1920s, and he had followed both Parks and Sternberg with senior positions at Canada's two leading natural history museums. Although he did notable work on dinosaurs, he was part of a new generation of palaeontologists for whom dinosaurs were only one of many interesting groups.

Loris Russell was born in New York in 1904, but his parents moved him to Calgary four years later. He spent a summer collecting with Charlie Sternberg in 1923, and took a B.Sc. at the University of Alberta in 1926. He then went to Princeton to study with William Berryman Scott

(1858–1947), a leading researcher on fossil mammals, completing an M.A. in 1929. In 1930 he completed a Ph.D., published his first dinosaur paper (a summary of North American Upper Cretaceous dinosaur faunas), and joined the Geological Survey of Canada as an assistant palaeontologist.

In 1937 (after the death of W.A. Parks), Russell became assistant professor at the University of Toronto, held in combination with the position of assistant director (with responsibility for vertebrate palaeontology) at what was then the Royal Ontario Museum of Palaeontology. During the Second World War Russell served from 1942 to 1945 with the Royal Canadian Corps of Signals (rising to the rank of major), and then became director of the same museum, and continued as an assistant professor at the University of Toronto.

Even before he had first graduated, Russell had published papers on fossil molluscs and mammals. He went on to publish on the geology of western Canada, fossil fish, Cretaceous and Tertiary mammals, and mammal tracks, and through his long career continued to publish on these and many other topics. A steady stream of papers showed his continuing interest in dinosaurs; he explored the little, mystery carnivore *Troodon*, and other obscure dinosaurs, and described the muscles of ceratopsians, bones associated with the eyes of duckbills, and the long crest of the remarkable duckbill *Parasaurolophus.* He also began to look at wider issues, such as the Cretaceous-Tertiary transition, when the dinosaurs disappeared, and changes in land and sea during the Cretaceous period.

Alberta Vertebrate Palaeontology Conference participants.

BATTERING RAM CERATOPSIANS

In 1950, Loris Russell became chief of the Zoology Section at the National Museum, and in 1956, director of its Natural History Branch. Two years later, he added the duties of acting director of the Human History Branch of the museum, and held the shared responsibilities for five years before moving back to the Royal Ontario Museum. During this period Russell became very involved in the developing museum movement in Canada, earning (with distinction) the Museums Diploma of the Museums Association (in Britain), and producing many papers on museum matters, and a couple of books on historical topics.

When Charlie Sternberg retired in 1950, he was replaced in 1954 by Wann Langston, Jr., recently described by a colleague as a "crusty Texan." Son of a medical doctor, Langston had spent time as a child in Vienna, where he saw the cast skeleton of the great sauropod *Diplodocus* every day. He went to the University of Oklahoma and then to California, and studied fossil crocodiles for his Ph.D. About the time Charlie Sternberg was retiring, Loris Russell visited Berkeley. There he met Langston, and was impressed by his knowledge of Canadian dinosaurs. When he joined the staff at the museum, Langston had never been to Ottawa, and (arriving on a rainy October day) found it more of a climate shock than a culture shock. However, he enjoyed the company and knowledge of his predecessor: Charlie Sternberg was "about the pleasantest old gentleman; just a gem." Langston worked on many kinds of Canadian vertebrate fossils. In 1955 he made a reconnaissance of the western localities with Russell, rediscovering the bone bed at Scabby Butte. He collected a duckbill in 1956, but was particularly intrigued by what he called the "battering ram ceratopsian," *Pachyrhinosaurus.* Fragmentary skulls of "this bizarre and enigmatic dinosaur" had been found there previously by Charlie Sternberg. In 1957 Langston returned to Scabby Butte, and from a quarry 6 x 15 metres (20 x 50 feet) recovered more than two hundred bones of dinosaurs — the first serious study of Canada's bone beds as a source of data. Work at Scabby Butte was continued in later years, and trips were also taken to Weston's locality at Irvine Coulee, other western sites, and Prince Edward Island. During his last season with the museum in 1962 Langston collected a *Triceratops* from southern Saskatchewan.

Langston published a number of dinosaur papers while in Canada. After he moved to Texas in 1963 he continued to publish on Canadian material, including papers on *Pachyrhinosaurus*. Some of his research was based on a new skull collected near Drumheller, which was placed on display in the Drumheller Fossil Museum. The Scabby Butte fauna was described in 1974 and 1975, and shown to include fish, amphibians, and mammals, while the reptiles included turtles, crocodiles, and a marine mosasaur, as well as dinosaurs. Langston checked the rock filling inside a *Pachyrhinosaurus* skull for pollen and spores, finding evidence of twenty-eight types of plants and learning more about the palaeoecology of the site. This sophisticated approach is an indication of how fast dinosaur studies were changing.

SKELETONS, BONE BEDS, AND MICROVERTEBRATES

The first collectors took any bones they could find and collect. Once techniques were available, partial and complete skeletons became the first choice. Dinosaurs previously described from isolated bones could be linked to more complete discoveries, and good display specimens helped to interest the public. Many skeletons are "headless wonders," with missing skulls. This suggests the carcass has floated after death for a long enough period for the bones and ligaments holding the skull to the body to decay. Skulls made up of solid bones may be found separately; weaker ones broke apart and became lost.

When skeletons were scattered by carnivores and scavengers and washed into rivers after death, the bones sometimes accumulated in bone beds, deposited downstream from the death site. Some bone beds seem to have accumulated over a period of time and include bones of several kinds of dinosaurs, while others are formed by a single flood event and contain the remains of a single species, suggesting some kind of disaster to a herd.

Bones of small dinosaurs, crocodiles, turtles, and even smaller mammals, snakes, and amphibians are called microvertebrates. They may be concentrated into beds of small bones by currents, and are sometimes found by scrutiny of pebbly layers, or even exposed by the diggings of ants. Some occur in little pockets that have been interpreted as coprolites (dung) of predators. Although a few small bones have usually been picked up by keen-eyed collectors, serious study of this material was not possible until methods of bulk processing were developed, concentrating bones and teeth.

"MR. PICKWICK" AND DR. RUSSELL

One of the most distinguished dinosaur specialists at the Alberta conference was Dr. William Elgin Swinton (1900–1994). Born in Kirkaldy, Scotland, he joined the staff of the Natural History Museum in London in 1924, and became one of Britain's few vertebrate palaeontologists, a historian of palaeontology, and a museum professional of broad sympathies. He was always enthusiastic about education; his activities included giving lessons in natural history to Queen Elizabeth II when she was a girl. At the museum his collection included the Alberta armoured dinosaur *Scolosaurus* collected by Cutler and described by Nopcsa, and he published an article on it as early as 1927. During the Second World War he spent time in Naval Intelligence (where he supervised Ian Fleming, the future creator of the fictional spy James Bond).

About 1958, I heard Swinton lecture on dinosaurs when I was an undergraduate at the University of Sheffield, England, and later studied for the Museums Diploma he had established. He had been booked to lecture in Alberta during the 1963 conference, but ended up mired in a swamp where he passed out and had to be rescued by helicopter. So as not to disappoint the audience, Colbert and Loris Russell delivered an impromptu lecture using Swinton's slides.

A bachelor, Swinton had lived in London with his mother, whose death eased the transition when in 1961 he came to Canada as head of the Life Sciences Division of the Royal Ontario Museum (formed when the separate subject museums amalgamated into one institution) and a professor at the University of Toronto. Two years later he became director of the ROM, with dramatic effect. He was a popular speaker, and is remembered as having "a slight Scottish burr and the jolly appearance of a Mr Pickwick, [which] dispelled the natural apprehension a lay audience might have felt on being addressed by a scholar." This talent was used effectively in the new medium of television, and perhaps as a result, attendance and government funding both increased, along with the academic stature of the institution.

Despite considerable pressures on his time, Swinton did not forget dinosaurs. He wrote a popular leaflet on Canadian dinosaurs for the museum, updated his pioneer book, *The Dinosaurs* (of which a new edition was published in 1970), and contributed work on dinosaur brains. Swinton retired from the museum in 1966, becoming Centennial Professor of the History of Science and helping with the planning of the innovative Ontario Science Centre.

A vacancy was created by Swinton's promotion in 1963. Loris Russell returned to the Royal Ontario Museum as Head of Life Sciences (a position soon renamed Chief Biologist), and also served as a professor in the Department of Geology of the University of Toronto. On tour to become acquainted with Canadian museums, I visited him in 1968 (when we were perhaps the only two in Canadian museums with the English museum qualification). Silver-haired and with a small military moustache, he soon made me feel at home and was ready to share his knowledge. From these positions Russell retired as Professor Emeritus in 1970 and Curator Emeritus in 1971, but continued to make annual field expeditions to the age of eighty-three. His flow of papers (including ongoing work on dinosaurs) continued almost to his death in 1998.

Some of Russell's later dinosaur work will be mentioned in appropriate places in later chapters, but this is perhaps the best place to mention his brief but remarkably complete summary, *Dinosaur Hunting in Western Canada,* published in 1966. This was the first coherent account of dinosaur collecting in western Canada — and a direct inspiration for the book you are now reading.

"I JUST STAYED WITH DINOSAURS"

"As a child, I could imagine the world of dinosaurs. At times I was a dinosaur. And so rather than do something practical when I grew up, I just stayed with dinosaurs." With this early beginning it is not surprising that Wann Langston's successor at the National Museum was ready to make dinosaurs a priority, yet his early training had led in quite different directions. American-born palaeontologist Dale Russell (who is no relation to Loris Russell) came to Canada in 1965. Born in California, Dale did a master's degree at the University of California (on Tertiary mammals). His Ph.D. was with Colbert — by this time the dean of dinosaur specialists — at Columbia. Yet, as his collaborator, Peter Dodson, comments, "In the mid-1960s dinosaur research was a very quiet field indeed," and Dale's dissertation was a study of mosasaurs, a group of marine reptiles infrequently found in Canada.

"Then as now," comments Dodson cynically, "museums unfailingly select curators who have not trained in dinosaur research over those who have." Whatever its agenda, the National Museum selected Russell. Unlike Langston, he did not make a quick return to the U.S., but set out to make an impact on Canadian dinosaur studies. "Tall, loquacious, witty and manically imaginative . . ." says Peter Dodson, "He threw himself into the study of Canadian dinosaurs with passion and abandon, and a steady stream of important dinosaur papers was soon emanating once again from Ottawa." Over the next thirty years (during which the museum went through many upheavals, changing its name to the National Museum of Natural Sciences and then to the Canadian Museum of Nature) Russell continued his energetic and innovative work on Canadian dinosaurs. Only in 1995 did he return to the U.S., taking positions at North Carolina State University and North Carolina State Museum.

Among his many interests, he has worked on carnivorous dinosaurs, with particular attention to the neglected smaller forms. His exploration of the evolution of intelligence has involved him as an adviser with NASA and the European Space Agency. He has taken an active interest in the hot-blooded controversy, paying particular attention to the problems of reconstructing the palaeoecology of the Dinosaur Park area. He has been Canada's principal proponent of the catastrophic extinction theory, and has convened two K-TEC (Cretaceous-Tertiary Environmental Change) conferences to discuss the issues. He has taken a strong interest in dinosaurs of other parts of the world, visiting most dinosaur-producing areas around the globe. He has also been Canada's leading popularizer of dinosaurs, commissioning many paintings and producing two illustrated books, *A Vanished World* and *An Odyssey in Time.* Some of these projects are discussed in later chapters.

Dale Russell photographs a find.

AN EVEN RARER TREASURE

Professors from both the geology and zoology departments at the University of Alberta were sponsors of the 1963 conference, but since Allan's retirement little dinosaur work had been done. Charles Stelck (a specialist in invertebrate fossils and stratigraphy trained by Charles MacLearn, discoverer of the Peace River trackways) had collected some material, such as a block of eight dinosaur footprints collected from the Pouce Coupe River in 1951. The 1963 conference provided enough leverage for a new position to be created for a vertebrate palaeontologist, a joint appointment in zoology and geology. In 1965 the position was filled by Richard Fox, an American who had just completed a Ph.D. at the University of Kansas. His research was on the cutting edge in both content and technique, for his focus was not dinosaurs; he "targeted an even rarer treasure: fossil mammals of dinosaur age."

Lambe had reported on Cretaceous mammal finds early in the century, but (compared with the dinosaurs) their teeth and bones were tiny, and they were rarely observed. Although chance finds had been seized on eagerly, little had been done to search for them until Loris Russell and Wann Langston began to look systematically for microvertebrates, using the new technique of washing quantities of loose rock to concentrate the bones. Fox was the first to use these techniques on a large scale in Alberta (commencing with a joint program with the University of Kansas), and he soon discovered an amazing variety of tiny mammals that were contemporaries of the dinosaurs. After a couple of decades of intensive work he was able to report that "No other part of the world has yielded specimens of Late Cretaceous mammals comparable to those from Alberta and Saskatchewan."

In 1970 the university's Laboratory for Vertebrate Paleontology (UALVP) was established. By 1988 its collection had twenty-four thousand study specimens, many of them tiny teeth stuck on the ends of pins. Fox has also collected some dinosaur material, such as hadrosaurs and *Centrosaurus* at Sandy Point on the South Saskatchewan River, and attacked the logistical problems of extracting heavy material from the badlands by getting a lift from a Canadian Armed Forces helicopter. UALVP has attracted visiting scholars from many institutions and has been a focus for a growing number of graduate students, some of whom have worked on dinosaurian problems. Such names as Paul Johnston and Bruce Naylor of the Royal Tyrrell Museum staff; Hans-Dieter Sues, now of the Royal Ontario Museum; and Peter Dodson, now of the University of Pennsylvania, all owe an important part of their training to the University of Alberta. Despite the cautious approach suggested at the 1963 conference, UALVP has continued to grow in importance, and was enhanced in due course by Mark V.H. Wilson, a specialist on the evolution of fish.

Fox and his students broke new ground in North America with dinosaur research as well as mammal study. American-born Peter Dodson was raised in Canada, and took his first degree in Ottawa. For his master's research in the late 1960s he applied the relatively new discipline of taphonomy to Dinosaur Park. Taphonomy looks at what has happened to animals since the time of their death. Dodson examined the quarries marked by Charlie Sternberg, and looked at the remains of the stream channels and other formations that were their sedimentary environments. The larger dinosaurs in the main streams were often oriented east-west, in the direction that the ancient water flowed, while those in side channels were across the current directions and (in the lesser currents) better preserved. This valuable study launched Dodson into dinosaur work, and also provided a basis for later investigations of the park area by a new provincial institution.

NO DIGGING!

When the Provincial Museum and Archives of Alberta was in planning, the minister in charge suggested that all items needed for display could be borrowed from Calgary's Glenbow Institute, and instructed the museum consultant that "there'll be no digging," thus brushing aside both archaeology and palaeontology. Fortunately, this demand fell by the wayside as the institution gradually developed, and the museum later developed notable programs in both fields, which have expanded even more in the hands of the separate organizations that later evolved from the museum.

The museum consultant was Ray Harrison, an Australian-born architect who was extension supervisor of the Saskatchewan Natural History Museum when he became a planner for the proposed Alberta provincial museum in 1962. Harrison attended the 1963 conference, and envisaged "the establishment of a paleontological program in a provincial museum, with a system of local museum staff employed in the field, conveniently located for getting to the materials." With Canada's centennial in 1967, the project was able to attract both federal and provincial funding, and Harrison's plan was far more ambitious than the province had originally anticipated. He eventually became the first director of the institution.

Recognizing his lack of subject and museum expertise, Harrison hired two head curators. The first, his former colleague Bruce McCorquodale (1924–1997), had also attended the 1963 palaeontology conference as Curator of Paleontology at the Natural History Museum in Regina. In 1964 he moved to Alberta as Head Curator of Natural History, later was transferred to the equivalent position in Human History, and still later became director of the museum. McCorquodale hired Don Taylor to begin an earth science program. Taylor had been an oil geologist in Venezuela, and later a technician at the University of Alberta; he began collecting bones for the museum in 1966. He was a dry humourist; a typical remark was that the cafeteria had remembered to put artificial colouring in the coffee, but forgotten artificial flavouring.

In 1967 I came to Alberta from England to replace McCorquodale and build the museum's natural history program. When I arrived in October, the museum building was nearly finished, and the initial temporary exhibits almost completed for the opening in December. The new $8 million facility was at that time by far the most spectacular and modern museum building I (or the province) had seen. Built of honey-coloured, fossil-rich Tyndall stone from Manitoba, it featured bronze sculptures outside and huge, empty spaces inside. The sculptures on the natural history side of the building featured Alberta dinosaurs and other extinct animals; I was told that the sculptor initially had featured *Stegosaurus* (from the U.S. Jurassic) instead of Alberta's *Stegoceras,* and that a fair amount of pressure had to be brought to bear to get it changed.

It was the prospect of real dinosaurs that excited me. When McCorquodale first showed me around the building, we entered a huge empty basement storeroom with three packing cases in the centre of the floor. These, I learned, contained a complete *Lambeosaurus,* collected by the National Museum in 1937 (the year I was born), and had never been unpacked. It had been collected by Charlie Sternberg, and had been named after Lambe by Parks; this aroused my interest in these dinosaur researchers, an interest that has led ultimately to this and other books. One of my first jobs was to supervise the preparation of this neglected dinosaur. Neither I nor technician John Poikans (who had been — and later became again — a house painter) had ever done such a job before, but I had some experience with other vertebrate fossils, and the resulting panel mount was successful. It was on show in Edmonton for a number of years, and is now in the Royal Tyrrell Museum.

The *Lambeosaurus* skeleton anchored the gallery, and was joined by two casts of life-size models of *Ankylosaurus* and *Corythosaurus* made by taxidermist/sculptor Louis Paul Jonas for the 1964–65 New

York World's Fair. A miniature diorama of a Cretaceous scene was constructed by artist Ralph Carson, and with some other specimens, a basic dinosaur gallery was created, and systematic additions kept it up to date with new discoveries.

DANGEROUS DIGS

Don Taylor was enthusiastic but mainly was finding isolated bones. We recognized the importance of dinosaurs for public display but knew that many skeletons had been shipped out of province. Were there, we wondered, any complete dinosaurs left to collect? Taylor located a partial carnivore (we called it *Gorgosaurus)* in the Manyberries area, which we collected. This was my first dig, in a dry coulee where the bones of the extinct reptile were guarded by a remarkable number of living reptiles — the first rattlesnakes I had seen. I was fascinated by the task of chiselling the hard rock around the fragile bones. Quite as unfamiliar was the tiny hotel in the small town of Hilda, where the locals put salt in their beer, and the bar fights spread along the corridor at night. Although these conditions were commonplace to earlier dinosaur hunters, I was still fresh from England, and the whole experience had an exotic air.

It seemed that more bodies might help. In those heady days, one wrote new budgets that were bigger than the present ones, and the additional cash often became available. With additional positions, Taylor could specialize in geology, and I was able to look for an experienced vertebrate palaeontologist. John Storer, who had done his Ph.D. with Loris Russell at the University of Toronto, joined the museum in 1970. Although his personal research interests were in fossil mammals (and he has continued to do important work in this field, particularly in Saskatchewan), Storer was ready to deal with dinosaurs. He rapidly got a handle on the stratigraphy, geology, and palaeontology of Alberta's Upper Cretaceous, and began to bring in partial skeletons. We were all a little inexperienced, and learned on the job. My most vivid recollection of fieldwork with Storer was the occasion we had to retrieve a plastered bone, too heavy to lift, which was perched halfway up a steepish hill. By now we had a four-wheel-drive truck equipped with a winch, but (as the quarry was on top of a hill) we could not get this above the site. The only way to remove it was for some of us to pull it down with ropes, while colleagues tried to steady it with other ropes from above. It was too heavy, and those of us who were pulling had to run to get out of the way — the only time I've ever been chased by a dinosaur.

Gorgosaurus *emerges.*

Storer began a series of publications in 1974 with *A Guide to Alberta Vertebrate Fossils from the Age of Dinosaurs*, an illustrated bone-identification manual featuring Hope Johnson's drawings. After making it clear that important dinosaur fossils were still to be found, Storer moved on to Saskatchewan, to be nearer to his beloved Wood Mountain mammals, and is now in Yukon.

"DOING THINGS WITH THEIR HANDS"

"In a less opulent society there were people who, from sheer love of doing things with their hands, went into the technical aspect of Vertebrate Palaeontology as preparators and developers." Wann Langston's comment about technical support amounted to the old complaint "you can't get good help" — or at any rate, at a salary you are able to pay. Certainly, in earlier years Canadian institutions neglected their curators, paying them poorly and expecting them to do their own technical work, and the only technicians were dedicated people, willing to do the job for the love of it, and not for much in the way of financial reward.

Langston recognized that "Alberta may be in a somewhat key position . . . because of the largely amateur interest that makes it possible to find such persons." And certainly, in my experience at the Provincial Museum, some excellent technicians — both men and women — could be found — though it did no harm that we were able to pay reasonable salaries.

Curators tend to get what kudos there are in the dinosaur business — they are the ones who describe the new species and appear on television talk shows. But every curator knows his success depends on his technical staff, and it has been my privilege to know and work with many fine technicians.

In palaeontology, technical work demands both physical endurance and enormous patience. Technicians have to live in field conditions in the summer, undertaking hard physical work in the hot sun or wrangling skidding vehicles on greasy roads in rain storms. Skill is needed to find and dig bones, but also needed are the skills to lift heavy objects safely, train and supervise volunteers, and obtain, manage, and sometimes function without vital supplies.

Once back in the lab, plaster jackets are opened and bones laboriously exposed, using noisy equipment and smelly — sometimes toxic — chemicals. Weeks, even months, may be required to prepare a single specimen, which may be so delicate that a slip of a tool could destroy vital evidence.

The main reward is pride in the job — the fascination of following bones through the rock. The technician's work is rarely acknowledged on the label on a specimen on display, though it may receive a credit in the fine print of a scientific description.

HADROSAURS IN HERDS

While the Provincial Museum was seeking to replace John Storer in 1975, an unexpected crisis erupted. John Sulek, a Calgary oilman back from a fishing trip, phoned. He had seen the Peace River dinosaur tracks (which we were aware of but had never seen) and gave us disturbing news that they would soon be flooded by a B.C. Hydro dam. This was — literally — out of our province, but when we talked to the British Columbia Provincial Museum we found that they had no palaeontology program at all. After some fast phoning, we had encouragement from the B.C. Museum, permission from the B.C.

Archaeological Survey, a grant from Sulek's company, Imperial Oil, and approval from our own government to work out of province.

We knew that since Charlie Sternberg's work in the 1930s the only fieldwork that had been done on the Peace River tracks had been a 1965 trip led by Gordon Edmund of the ROM, which had made moulds of two trackways found when the W.A.C. Bennett Dam was being built between 1961 and 1968. Now a new dam lower down the canyon had been started, which would inundate the whole canyon by 1979.

We assembled some staff: Curator of Geology Ron Mussieux and technicians Linda Strong and Ron Solkoski. Our freshly hired curator of palaeontology, Phil Currie, was ready to take charge. Since Currie had no experience with footprints, I also asked Bill Sarjeant, Professor of Geological Sciences at the University of Saskatchewan, to participate. Sarjeant had come to Canada in 1972 from England, where he had done important work in fossil reptile footprints, so at this point he was the only western Canadian with any expertise in this then-unfamiliar area.

"The first day, we drove out to the Peace River Canyon," recalls Sarjeant. "The water was high, and we were pessimistic about finding anything." Currie's response was to find and hire Carl Kortmeyer of Dawson Creek, whose powerful jet boat and personal enthusiasm — later supplemented by his son's — became essential to continuing work in the canyon. The boat provided a safe but exciting ride through the violent rapids (including four-foot standing waves) that had previously been regarded as unnavigable. Kortmeyer was able to carry personnel, equipment and materials in, and casts and excavated footprints out.

"We went up the river," Sarjeant continues, "and found an outcrop . . . with what appeared to be the edge of a footprint showing from underneath a thin layer of argillaceous rock. . . . we began to strip this back and were soon finding more and more footprints — ultimately we uncovered a surface 20 feet by 8 feet with many crisscrossing dinosaur tracks. We were thrilled to pieces." A charred log lying on the shore was used to outline the tracks so that they could be photographed. Next day, further up the river, a much larger footprint was found at a river-edge outcrop. Rainwater had accumulated in it and had to be sponged out so that the footprint could be cast. The original bedding plane proved to have more than 150 hadrosaur footprints of what proved to be a new type, and other tracks were everywhere along the river's edge.

It was clear that there were far more footprints than Sternberg had seen, and that one expedition could not discover all of the significant footprints, let alone rescue them. During the next three summers, Provincial Museum expeditions returned; B.C. Hydro contributed both financially and by regulating water levels in the river from the upper dam. Assistance was provided by Ross Brand of the B.C. Provincial Museum and Charles Stelck and Jeff Doran of the University of Alberta; Doran joined the museum staff before the end of the project.

The whole thing had been, as Dodson comments, "a marvel of ingenuity and endurance under truly trying conditions." After four years, ninety actual tracks had been cut out with rock saws and collected, while rubber moulds had been made of another two hundred. Over one thousand more footprints, in more than one hundred trackways, had been mapped and measured. What proved to be the richest dinosaur footprint site in Canada had been adequately sampled, but only partially rescued.

Research based on the recovered material continued for some years. Hadrosaur tracks were most common, including rare handprints and tracks of baby hadrosaurs. It was not possible to relocate the original type specimen of Sternberg's *Amblydactylus gethingi* until 1979, "when it was spotted underwater from a helicopter." Sarjeant and Currie jointly described a new *Amblydactylus* species, named *kortmeyeri* after its discoverer Carl Kortmeyer.

> The main significance of the Peace River Canyon footprints is the fact that they occur in *trackways*. . . . They show that hadrosaurs traveled in herds, and suggest that the animals walked side by side rather than in clumps. Baby hadrosaurs appear to have walked in groups of animals of the same size. Many of the footprints were made by swimming animals just touching bottom. The large carnivorous dinosaurs seem to have hunted in small packs, and appear to have been equally at home in shallow water and on the shoreline. Small carnivores appear to have traveled in groups as well. None of the trackways indicates high activity levels, although the carnivores are obviously more agile and quick than the herbivores.

One of the most exciting discoveries was made by Currie and photographer Suzanne Swibold. "As the final days closed on the expedition," remembers Currie, "we stopped for the last time at a familiar locality. We thought the area had given up all its secrets, but we were wrong. The river had unexpectedly risen overnight, and as the water fell away again, a rock layer split to reveal more than 150 small footprints. I immediately realized they were bird footprints." At the time these were the earliest record of fossil bird footprints from anywhere in the world.

Hadrosaur *vertebrae, Dinosaur Provincial Park.*

Meanwhile, Rick Kool of the B.C. Provincial Museum had noticed an odd track in the Hudson's Hope museum. He mentioned this to Sarjeant at the Regina Folk Festival (where by coincidence they were both performing) and later sent him a cast. After laborious research, it proved to be the oldest track of a marsupial from anywhere in the world, and the oldest mammal track in North America.

With hindsight, it may seem strange that such a major development as a new dam could go ahead without any investigation of known and important fossil resources. However, these were the early days of environmental-impact studies, and they were not required on most projects. Alas, even four seasons were not enough. Currie regretfully reported: "At the time of flooding there were at least several hundred more footprints that we did not have time to document, and stretches of the canyon that had never been searched. A third type of footprint as yet unnamed was recovered in the last week of the last expedition." My own field notes near the end of the first season (well before we knew the full extent of the resource) noted: "Shame to have to take out these fine footprints. This area should have been a park, with the whole bed exposed and protected." By the 1980s, such a resource would have been seen as having major tourist potential.

PRINCE VALIANT HUNTS DRAGONS

As a boy, Phil Currie persuaded his parents to take him to visit western Canada so that he could see dinosaurs. He was deeply disappointed that there weren't many to see. A few decades later, Currie has done more than anyone else to change that deplorable situation. When plunged into a logistically, politically, and scientifically complex situation in the Peace River Canyon, Currie acted as if he had been waiting for years for just such a chance, and knew just what to do.

Currie became aware of dinosaurs at the age of six when a plastic fossil model fell out of his cereal box. Living in Toronto, he was able to spend time in the dinosaur galleries at the Royal Ontario Museum. He discovered Roy Chapman Andrews' books about dinosaur hunting in the Gobi Desert, and was enthralled by the murals of dinosaurs at the Field Museum in Chicago. Fictional dinosaurs also appealed; Currie was a fan of Edgar Rice Burroughs, and for a few years published his own fanzine, *ERBivore.* He set out to become a vertebrate palaeontologist, completing his first degree in Toronto, taking as many courses as he could from Loris Russell, and following with a master's thesis at McGill with Robert Carroll. He was in the middle of his Ph.D. thesis (on a very obscure Triassic reptile from Madagascar) when the Provincial Museum vacancy in Alberta was advertised in 1976. Currie recognized that this was an unusual opportunity, put his Ph.D. to one side for a while — and was hired.

Twenty years after Currie's arrival, Peter Dodson hails his move to Alberta as "a most important development," when "a lanky, long-legged, curly-haired young man . . . packed up his papers and moved to Edmonton." The female education staff were equally impressed, and promptly nicknamed Currie "Prince Valiant," after the cartoon character. In Alberta, Prince Valiant rode off into the wilderness in search of dragons to such good effect that the dinosaurs almost galloped into the museum. In addition to the Peace River rescue, Currie worked on a number of other important dinosaur projects.

Partial skeletons of duckbills were collected in 1976 and 1977, and more duckbills and horned dinosaurs in 1978 and 1979. A growing network of contacts brought in new material from around the province, the most important of which (such as the pachycephalosaurid *Gravitholus* and the tiny carnivore *Saurornitholestes)* were in due course described by graduate students at the university.

During the early years of the museum, senior management of the Provincial Parks Department had been rather cautious about museum involvement in Dinosaur Park, but by the end of the 1970s new staff and new attitudes in head office — coupled with Currie's open friendliness — made it possible for the museum and park to begin close cooperation. In 1979 a lambeosaurine skeleton was collected near the entrance of the park, and a number of exciting finds were made of small carnivores. One of these was made by Gilles Danis, who had moved from the National Museum to the Provincial Museum of Alberta as a senior technician, and who became the core of what Dodson refers to as "a first-class team of dedicated technicians."

In the same year, 1979, a systematic survey of the park documented more than one thousand isolated finds, including a pterosaur, skullcaps of the boneheaded *Stegoceras,* and microvertebrate sites. One of Currie's innovative approaches was systematic study of bone beds, a process barely started by Wann Langston at Scabby Butte. Many bone beds were surveyed in the park, and in 1979 Currie started a detailed study of one dominated by remains of the horned dinosaur *Centrosaurus.* The bone bed became a prime attraction for visitors to the park.

By the early 1980s, palaeontology staff was up to two curators and four technicians, and these were joined by temporary staff and volunteers, bringing numbers as high as eighteen at a time. In subsequent years, Currie was able to put together an interdisciplinary, interinstitutional team to undertake an integrated study of Dinosaur Park, which continues to this day.

Dinosaur Provincial Park as seen from the lookout.

Currie not only was a researcher, but also used his flair for education and promotion to the full. For Alberta's seventy-fifth anniversary in 1980, a temporary exhibit, "Discovering Dinosaurs," was opened, with material collected by the program and borrowed from other institutions in Canada and the U.S. It attracted thirty thousand visitors in its first week, and broke all attendance records for the museum. A visiting scientist called this temporary exhibit "one of the ten best on dinosaurs in North America." An associated program showed classic movies featuring dinosaurs, and brought in dinosaur scholars such as Robert Bakker, Peter Dodson, John Ostrom, and Dale Russell to speak. Other displays were produced at Dinosaur Park for the World Heritage Site dedication, one showing footprints at the Peace River Canyon.

DINNER WITH THE DINOSAURS

The invitation's cover showed an *Albertosaurus* in a dress coat, bowing and doffing his top hat. This was not a usual invitation. "We collectors of dinosaurs cordially request your formal presence at a formal evening of camaraderie in the badlands of the South Saskatchewan River. The place is the Sandy Point Crossing and the time the evening of Saturday, August 26, 1978." RSVP was to Philip J. Currie at General Delivery, Medicine Hat, and a map on the back of the invitation pinpointed the Sandy Point Crossing, close to the Saskatchewan border where Alberta's Highway 41 crosses the South Saskatchewan River.

Dinosaur dinners were another innovation by Currie. An end-of-field-season festivity in an appropriate spot in the badlands, the dinner was a celebration of the end of the field season, a thank you to volunteers and other helpers, and a rendezvous for other researchers who were in the province during the summer.

It was a long, hot drive through golden fields with harvest in full swing. As we reached Sandy Point, the blue skies turned grey for the first time. The camp was on the south side of the river: a trailer borrowed from the Highways Department, and a ragged string of tents and trucks. We went to see an old dig site, and picked up eroded rocks, which (with candles) made splendid centrepieces for the table. Tarps and polythene sheets were jury-rigged over the tables in case of rain, tied to the trailer at one end and the cottonwoods at the other, supported with string, 2 x 4s, and ad hoc tent pegs. Good job it didn't rain.

There was quite a crowd: all of the field staff, some with wives and kids and dogs, and volunteers. Staff were visiting from the university and Dinosaur Park. Former curator John Storer was there from Regina, and Loris (and Grace) Russell from the ROM. Distant visitors included George Olshevsky (a dinosaur enthusiast from California who promptly christened me Captain Spa[u]lding after Groucho Marx's character), a geology student from Utah, and a graduate student in vertebrate palaeontology from Scotland.

Dress was formal — long dresses for the women, and jackets and ties for the men. Above the table all was elegant, but the effect often went only as far as the waist or ankle, below which jeans and sneakers could be seen. The children were fed first (hamburgers and ice cream), while we older folk began cocktails (red rotgut in plastic glasses) at seven. The preliminaries continued until 8:30, as expected guests were still arriving. People chatted, and photographed each other in the gradually failing light. Killdeer called by the river, and occasional cars drove by with faces staring out, baffled by such a large gathering in such a remote area.

Soon dinner was served, with younger members of staff and volunteers (one resplendent in white shirt and black velvet dinner jacket) handing round the lobster bisque. Official guests were at a formal high table, while the rest of us squeezed in where there was room. Coq au vin was followed by ice cream with crème de menthe, and then a short informal speech of thanks by Russell (looking like an ambassador) ended the formal evening.

The rest of the evening was spent around the campfire, with songs, jokes, and chat. "Retired exhausted, very late," says my field journal.

The next morning, we all had to pay the piper. Currie gleefully gathered a crew of able-bodied but somewhat hungover helpers. A big chunk of the frill of a horned dinosaur had been plastered and was ready to be loaded. The only snag was that it lay two-thirds of the way down a hill, and trucks could only be brought to the top. The four-wheel drive could safely get part way down the hill, but the other truck had the winch. So we attached a rope to the block and, with much grunting and swearing, hauled it up the hill over cactus and loose boulders until the winch cable could take over. A memorable occasion.

HOT BLOOD, HOT TIMES

In 1965 Loris Russell published a short paper exploring the possibility that dinosaurs were warm-blooded, but that their absence of a protective covering (unlike the birds) may have contributed to their extinction. "The silence was deafening," he commented later. Although the idea had come and gone since the last century, it had been quite overtaken by a more orthodox view of dinosaurs as traditional reptiles.

One of Colbert's students in the 1950s, John Ostrom, had also raised the possibility that dinosaurs had a mammal-like physiology. In 1964 he and a colleague found a new small carnivore, *Deinonychus,* which had an anatomy that suggested a capacity for swift and savage action. "It must have been," reported Ostrom in 1969, "a fleet-footed, highly predaceous, extremely agile and very active animal. . . ." Ostrom's student Robert Bakker took the idea and ran with it. His turning-point paper, "Dinosaur Renaissance," appeared in *Scientific American* in 1975, and argued confidently not only the warm-bloodedness of dinosaurs, but the novel concept that "the dinosaurs never died out completely. One group still lives. We call them birds." Bakker has been an outspoken (and persuasive) advocate for this idea ever since, and has been at the centre of vigorous and ongoing controversy.

At the Alberta dinosaur conference in 1963, the real enthusiasm of the academics was for neglected groups of vertebrates that had not yet received their share of funding. Dinosaurs were recognized by some to have their importance, but by others were being touted only as the popular key to the public purse. A decade later, dinosaurs were hot, hot, hot. Not only were the public and the media on the trail, but also a new generation of young academics that was anxious to find a foothold in the dinosaur field. Growing up like Currie with museum exhibits and true-life fossil adventures, they were quite convinced that dinosaurs were not passé. They wanted to re-evaluate previously collected material, find new specimens, and test new hypotheses. While they recognized the importance of other life — if only as an essential part of the world in which dinosaurs lived — many of them were keen to work on dinosaurs. All over North America (and to some extent elsewhere) a few dedicated individuals were scouring the ground, and making new discoveries.

WORLD HERITAGE SITE

In 1979, Dinosaur Provincial Park joined the growing ranks of World Heritage Sites, and in the summer of 1980 I attended the ceremony that celebrated its selection. The sun shone brilliantly, and the usually quiet prairie rim was a busy place. A helicopter stood to one side, in a roped-off landing pad, while a crowd of several hundred was watched over by Mounties, TV cameras, and two film crews. As they arrived, VIPs were presented with cowboy hats, while civil servants and the public at large greeted acquaintances and applauded the weather. The crowd gradually settled down on rows of chairs facing the platform, behind which the flags of Alberta, Canada, and the United Nations fluttered against the blue sky. Beyond spread the tumbled panorama of the park's badlands.

We were reminded that World Heritage Sites were established under a UNESCO convention that came into force in 1975, to "ensure, so far as is possible, the proper identification, protection, preservation and interpretation of the world's most significant cultural and national heritage." Canada's first sites were Nahanni National Park and the Viking settlement at L'Anse aux Meadows in Newfoundland. Dinosaur Provincial Park and Kluane National Park were the next to be established. Dinosaur Park was the world's first to be chosen primarily on the basis of its fossil resources, and (despite an unsuccessful bid by Parks Canada) was also the first not under national management. Flattering comparisons were made with other World Heritage Sites established to protect natural features, including Mount Everest, the Galápagos Islands, and the Grand Canyon of the Colorado.

A presentation to Douglas Martin honoured his grandfather, Charlie Sternberg (who was too elderly to travel). Finally, federal and provincial ministers together unveiled a plaque. In the flurry of self-congratulation, it seemed that the park itself might be forgotten, until (as Glenn Rollans remembers) "the man from UNESCO again retrieved it and set it before the crowd. He stood up, said a few words in Urdu, smiled, then translated. 'Where is paradise on earth?' he asked, opening his arms to the badlands. 'It is here, it is here, it is here.'"

As we contemplated the remote past in which the dinosaurs had lived, and the more recent century or so of discovery, none of us fully appreciated that this occasion was not just a celebration of achievement, but the birth of a new golden age for Canadian dinosaurs. Bakker's ideas were new and radical, and none of us fully realized that dinosaurs were on the verge of becoming one of the hottest scientific topics of the century. And although some of us were by then planning a new museum for Drumheller, we did not anticipate it would become one of the greatest fossil museums in the world. No one could have guessed that wonderful new finds would be made, not only in Dinosaur Park, but also in Canada from sea to sea — to sea. And no one would have dreamed that Canadian dinosaur scientists would soon be in central Asia, while Chinese scientists would be making discoveries in Canada.

7

"THE CROWN JEWEL OF PALAEONTOLOGICAL MUSEUMS"

"One of the great charms of palaeontology is that one never knows what will turn up where, or who will find it."

PETER DODSON

ONE NEVER KNOWS

In spring of 1983, Phil Currie and visiting colleague Jack Horner from Montana State University walked over the badlands in Midland Provincial Park, northwest of the city of Drumheller. Close by, the skeleton of a new museum building was rising from the badlands. They were chatting about Currie's interest in small carnivores when Horner "saw a fossil on the ground and stopped to pick it up. It was a piece of the jaw of a small carnivorous dinosaur. We knelt down to look at where it came from, and we could see that the rest of the jaw . . . was going into a hill."

Phil's crew planned to excavate the rest of the find, but "it rained. For a week. When we were able to get back there, it was gone. We couldn't find it." Horner rediscovered the specimen on another visit, two years later. It solved a 130-year-old riddle, for some of the teeth in the jaw were those that had been described as *Troodon*, while other teeth in the same jaw had been described as *Stenonychosaurus* and under a couple of other names. It not only proved that these names applied to the same animal, but set the course for later discoveries that would show the little raptor as close to the origin of birds.

This specimen, of great importance to science, probably never would have been found unless the museum in construction had brought the two palaeontologists to the site. It provided the first hint of the unexpected way in which the new facility would prove to be of importance in dinosaur research.

PALAEONTOLOGICAL MUSEUM AND RESEARCH INSTITUTE

When the press reported in 1979 that Minister of Culture Horst Schmid had promised that the province would build a new fossil museum in Drumheller, probably no one was more surprised than the Provincial Museum staff who were responsible for the province's palaeontology program. True, we were well aware of the energy and enthusiasm of the Drumheller museum group, and the associated boosters who created the Dinosaur Trail and the city's "dinosaur country" image, and who in 1978 requested $5 million of provincial funding for an enhanced fossil museum. There was also pressure on the Alberta government to share its oil bounty around the province, eventually leading to major public facilities of one sort or another in each major centre. A third pressure was undoubtedly the success of the Provincial Museum in finding dinosaurs,

Official opening of the Dinosaur Park Field Centre.

collecting them, and making them accessible to a larger public. Fourth, the senior staff of the government department responsible for heritage matters (Alberta Culture was its most common name) had developed an ambitious plan to place major interpretive facilities representing the important themes of Alberta. The network, which now includes such museums and interpretive centres as the Ukrainian Heritage Museum, Head-Smashed-In Buffalo Jump, and the Reynolds-Alberta Museum, clearly would be incomplete without proper attention to Alberta's most important palaeontological resource.

At first it was unclear what had been promised. From our ongoing relationship with the existing Drumheller fossil museum, we knew that their hopes were set on an extension to their building. But it soon became clear that what had been promised was a new facility, which would relocate the existing palaeontology program of the Provincial Museum, and perhaps expand it a bit. And all the planning was still to be done.

Phil Currie, Curator of Vertebrate Palaeontology at the Provincial Museum, was obviously a key person. Following preparation of a preliminary planning document in November 1980 for what what was initially called the Palaeontological Museum and Research Institution (PMRI), which envisaged a $14 million budget, I arranged for myself to be seconded to the project to work on interpretive planning, and Currie and I took over a rented space in the old Boardwalk building in downtown Edmonton. We spent 1981 in more detailed planning, by which time the budget had increased to $25 million. Existing palaeontology staff (including technicians Gilles Danis and Linda Strong) and a growing number of other support staff came in to undertake specific projects, including Brian Noble, whose imagination and drive eventually led the Canadian scientists to China.

Apart from the government's intention to put substantial dollars into the new fossil facility in the Drumheller area, almost everything was up in the air. A suitable site had to be located in the only provincial property available, Midland Provincial Park. Currie was trying to keep on top of his ongoing field programs, and we went after additional field staff knowing that we would need as much display material as possible. The minister's announcement notwithstanding, there was a key question, Would enough people go to a museum situated so far from a large city to justify its construction? We had to get planning studies under way, and develop cooperation with Parks, Public Works, and the other government departments involved. An academic advisory committee was set up, and meetings were held with Drumheller civic representatives. A building program had to be prepared, defining the spaces and their

Phil Currie.

characteristics, and an architect selected. I was particularly concerned with exhibit and program approaches: it was a time of fast development of museum display technique, and this museum would have to be as near to state-of-the-art as we could afford. I visited some key institutions in eastern Canada and the U.S., and identified many of the key elements of modern interpretation that were later used in the museum.

Gradually things fell into place. The park proved to be honeycombed from end to end with old coal mine workings, and only one site at the west end was both flat enough and solidly based. It was important to situate the building inconspicuously so that it would not dominate the badland landscape, which was an important part of the story we were to tell. The first facilities program was completed by July 1981, and the Bill Boucock Partnership of Calgary was hired as architects. Research was to be as important to the institution's mandate as display, so we needed attractive offices and workrooms and a good library — essential facilities to support and attract high quality staff to a relatively remote site.

My interpretive plan of December 1981 projected some of the highlights of what was possible. The PMRI was seen as potentially the first museum to integrate computers from the start, and one of the few to incorporate site interpretation in conjunction with major parks. Our experience at the Provincial Museum showed that the public was fascinated by the preparation of dinosaurs, so we planned to make the main lab visible from the public space. The exhibition of dinosaurs and other large fossils demanded drama, so we needed a space that could be laid out dramatically, with adequate height to allow changes of level and different viewpoints, and completely controlled lighting. To show the sort of plants that lived in dinosaur time we planned what became the palaeoconservatory, a giant greenhouse displaying modern plants whose relatives lived in Cretaceous times. Traditional means of painting realistic backgrounds to exhibits then involved a painter doing every brushstroke by hand — impossible on the scale and time frame we were working under. I discovered a then-new technology that allowed a smaller painting to be enlarged using digitization. Extensive use of conventional media — audio, film, and video — was projected. Computer-generated exhibits were also cutting edge for the first time, and we envisaged a number of different ways of using them interactively in the galleries — for instance, by showing the interrelationships of different parts of dinosaurs, in an educational game that might be called "Design-a-saur." Many of these features appeared in the museum's initial exhibits.

MOVE TO DRUMHELLER

The first director of the museum, David Baird (then director of the National Museum of Science and Technology) — not to be confused with American vertebrate palaeontologist Donald Baird — was hired in December 1981, and he arrived in Edmonton in January 1982. Within a few months he had moved with his staff to Drumheller, where temporary quarters were set up in a former supermarket. Phil Currie — in due course Assistant Director (Collections and Research) of the new museum — and the other palaeontological staff went, too, and the search for housing in the area began. Other staff (including Brian Noble and myself) left the project at that point, though both of us found other ways to continue our involvement with dinosaurs.

On May 15, 1982, a site dedication ceremony was presided over by Minister of Culture Mary LeMessurier. The museum was named the Tyrrell Museum of Palaeontology after Tyrrell, whose find less than five kilometres (three miles) away had brought attention to the dinosaurs of the Red Deer River. The museum's mandate was defined as "a celebration of three thousand million years of life on Earth with particular but not exclusive reference to Alberta." Detailed planning continued, and by the following year the building was rising from the badlands, and several design teams (both in-house and outside) were hard at work on its exhibits.

Dinosaur models outside the Tyrrell Museum.

HELPING SOLVE THE RIDDLES

Most major museums are in large cities, where there is a substantial base of population and a wealth of libraries, universities, and other cultural resources. Often such cities are remote from the major sources of fossils, which in turn are only acquired by lengthy expeditions, which are more or less disruptive to the ongoing research activities and family lives of the scientists involved. Drumheller is a small city, with a population in 1982 of 6500, and placing the Tyrrell Museum there was very much an experiment.

In general, the staff were enthusiastic, and most made the personal transition without much hardship. A major city, Calgary (with its university and Geological Survey of Canada office), was only 135 kilometres (84 miles) away, a trip generally manageable even during Alberta's cold winters. In due course, the Museum was able to set up cross appointments with the University of Calgary.

Without the need for endless treks to and from a home base in Edmonton, the research program seemed to take a great leap forward. Currie summed up the research mandate as he saw it: "The museum's great opportunity is to explore different problem areas in an unbiased fashion and discover the evidence for and against the theories of those who don't have the chance to work in the field. It's hard, exacting work, but the satisfaction of helping solve the riddles is immense."

A number of active programs were carried to the Tyrrell Museum from the Provincial Museum, but the staff was greatly increased. By 1982, the Tyrrell's first independent year, eighteen staff and fifty-four volunteers were in the field, in collaboration with three professional associates from other institutions. Five articulated dinosaur skeletons were among the three thousand catalogued specimens collected during that year; these were a continuing priority, of course, as the best of them would be needed for exhibits. Some came from systematic searches by museum staff of likely areas, others from environmental assessments now required by the province's heritage legislation. Still others came through chance finds by the public reported to the museum as its profile increased — for instance, three high school boys found a specimen in the Crowsnest Pass in 1981 that proved to be one of the world's few known specimens of *Tyrannosaurus rex.* Such was the success of discovery that there was always a backlog of articulated specimens waiting to be removed.

Methods of excavation were essentially unchanged, except for the larger crews that could be put on the job as staff and volunteers increased. However, a valuable innovation (starting when the program was still at the Provincial Museum) was the use of helicopters at the end of the field season to lift larger pieces of plastered bone from the badlands to a waiting truck.

Currie's interest in dinosaur footprints was by no means exhausted by his four seasons in British Columbia. "Using the knowledge gained from working on the footprints of the Peace River Canyon, staff of the Tyrrell Museum of Palaeontology have substantially increased the number of known footprint sites in western Canada. . . ." he reported in 1989. By this time he had twenty-seven documented sites, from latest Jurassic to latest Cretaceous, scattered from extreme southeastern British Columbia through central Alberta and the foothills in western Alberta and eastern B.C. Some of the best were from the ceilings of underground coal mines at Michel, B.C., which had been cast in the 1940s. An important trackway site was found close to Dinosaur Ridge, "a topographic ridge named for its sinuous shape" rather than for the tracks later found there. There, eight trackways contain more than two hundred footprints.

The Dinosaur Park cooperative project had been initiated in 1980 with cooperating scientists from several other agencies. Other dinosaur specialists included Dale Russell from the National Museum of Natural Sciences and Peter Dodson (who had done his master's thesis at the University of Alberta), and was now at the University of Pennsylvania. Stratigraphical work was undertaken by Emlyn Koster from the Research Council of Alberta (who became the second director of the Tyrrell Museum), and a variety of expertise came from the universities of Alberta, Calgary, and Saskatchewan.

Work continued on the *Centrosaurus* bone bed, and on many other projects. The park was also yielding much non-dinosaur material. In 1980, Dale Russell found the first evidence of the giant pterosaur *Quetzalcoatlus* — a flying reptile the size of a small plane — in Dinosaur Park. Microsites studied by washing quantities of loose sediment yielded bones of small reptiles, mammals, and other vertebrates, including bird remains from the park and other sites. Ongoing studies of fossil plants (including the fine pollen remains visible only through a microscope) were rounding out knowledge of the ecological context of the dinosaurs.

Fieldwork slowed during 1985, the year the museum was being fitted for its opening, but after that it resumed with great enthusiasm. Other areas of Alberta were also producing important material. In 1986, skulls of the "battering ram dinosaur," *Pachyrhinosaurus*, were located at Pipestone Creek, Grande Prairie, where another substantial dig was initiated, producing a new species now being described by Currie, Langston, and Darren Tanke. In the following year, the first dinosaur nests in Canada were found at Devil's Coulee, near the border.

Centrosaurus *bone bed, Dinosaur Provincial Park.*

QUARRY 143 — THE *CENTROSAURUS* BONE BED

At certain times of the year, large herds of *Centrosaurus,* a single-horned relation of *Triceratops,* browsed upon the open plains in the area that became Dinosaur Park. Adults and young of various ages travelled together, and a predator hiding in the vicinity would have heard their calls, and perhaps even their munching on coarse vegetation. Seeking an opportunity to strike, the predator may have watched the herd ford rivers. A disaster of epic proportions overtook this herd, and interpreting its cause has become a fascinating exercise in deduction.

Bone beds, in which the bones of individual dinosaurs are scattered, provide the only dinosaur material in many parts of the world, but had been long neglected in skeleton-rich Alberta. Some have been formed slowly and contain material from many different species, while others seem to have been formed quickly, and document some sort of natural disaster in which only one kind of dinosaur is involved.

Currie started to excavate a *Centrosaurus* bone bed (known as Quarry 143) in Dinosaur Park in 1979, and the project continued into the Tyrrell Museum years. By the end of the 1980 field season, only ten percent of the site had been explored, but eighty-five percent of the identifiable bones proved to be from a single kind of horned dinosaur. The bed, only a few inches thick, contained about twenty bones per square metre, though it was later found that some pockets were three times as rich.

Currie hit on a readily identifiable bone, the upper part of the eye socket that bears the base of a little horn. These are clearly identifiable as left and right, so it was possible to estimate the number of individuals represented by the collection (thirty-eight by 1980, fifty-seven by 1992). Three distinct size classes were recognized early, and did not have any real intermediates, suggesting that different ages were present in the herd. Assuming some seasonal variation and a definite breeding season, the youngest (two individuals) may have been six months old when they died, the next largest (five individuals), almost double the size, were perhaps yearlings. The rest were at least fifty percent bigger than the yearlings, and were presumably adults of different ages, growing less quickly than the young. Work on the same bone bed has continued into the 1990s, and it is now known to stretch for ten kilometres (six miles), and contain the remains of perhaps ten thousand individuals.

The death of a herd of the same kind of dinosaur suggested some sort of natural disaster. Currie initially considered epidemics, droughts, and floods, and soon found striking examples of mass deaths from flooding. For instance, in 1795 a fur trader witnessed the deaths of thousands of bison, in floodwaters of the Qu'Appelle River in what is now Saskatchewan. In 1984 Currie and Dodson published an interpretation of the bone bed as a flood disaster — yet the huge number of bodies was baffling.

After death, the bodies clearly lay exposed on a river bank or dry river bottom. Broken teeth and tooth-marked bones show that the bodies were well picked over by carnivores. The size of the bones remaining in the bone bed made it clear that water had then carried off the smaller bones, leaving the heavier ones to be covered up for seventy million years or so.

In 1980, another fifty bone beds were documented, one covering more than 40 hectares (100 acres), so there is plenty of further scope for research of this kind. A wider survey in 1984 by Currie and Dodson added monotypic bone beds elsewhere in Alberta. Dinosaur Park has bone beds of the horned dinosaurs *Centrosaurus, Chasmosaurus,* and *Styracosaurus,* and in the higher Edmonton Formation are others containing *Anchiceratops* and *Pachyrhinosaurus.* Patient research can tell us much more about the horned dinosaurs.

A GRACEFUL, LOW-SLUNG STRUCTURE

Meanwhile, the Tyrrell building was rising from the badlands; its cost eventually reached $28 million. It was designed in layers that pick up the horizontal bedding in the landscape, and lower layers were covered by berms to "promote the illusion of a building growing out of its environment." The access road leads off the Dinosaur Trail, and meanders among buttes before curving gracefully before the entrance and leading to parking beyond. Steps lead up the butte before the museum, and from its top, extensive views up and down the valley remind the visitor forcibly of the environment in which dinosaurs are found, if not that in which they lived. The building received a number of reactions, varying from "a graceful, low-slung structure, highlighted by plaza fountains," to a "low, squashed design and flat bentonite color."

Inside, the usual public facilities are grouped around the entrance, including an auditorium, a small shop (later expanded into the main hall), and a cafeteria. The main exhibit area was 3700 square metres (40,000 square feet), from 6 to 7 metres (20 to 24 feet) high, a "black box" in which lighting could be completely controlled. A full forty percent of the display area was set aside for dinosaurs and their environment in the Mesozoic. The palaeoconservatory is at one end of the displays; a greenhouse containing living plants is accessible from the galleries.

Apart from a window into the main preparation laboratory, the public does not see the working areas of the museum. These include the rest of the 11,200-square-metre (120,000-square-foot) building — more than equal in size to the displays — making it a very functional museum for the staff, who tend to be squeezed into odd holes and corners in older buildings. Extensive storage areas and labs below store the extensive resources of fossils awaiting treatment or not currently useful for exhibition. These are surmounted by the library, an essential resource that by 1990 had fifty thousand volumes, including twelve thousand books and fifteen thousand periodicals. Offices and design studios are arranged around the periphery, so that staff have views of the badlands, which contain their inspiration and their challenge.

Before the museum planning was even under way, tourists were arriving in Alberta and asking where to find it. By the time there was actually a museum to see, the September 15, 1985 opening attracted some five thousand people. They were followed by almost 600,000 more in the museum's first year, an astonishing attendance. Admission was initially free, though visitors were invited to contribute to the museum's research programs.

MEAT AND POTATOES OF THE TYRRELL

The visitors came, of course, not primarily to see the building, but the exhibits it contained. Three and a half years of detailed collection, design, planning, preparation, and construction had produced an impressive result. When the first visitors entered the building they were greeted by a "celebration of life," a last-minute addition to the exhibits in which 196 backlit photographs, flicking on and off, present a vivid series of images of living wildlife, perhaps a puzzling introduction to the visitor expecting a museum featuring fossils. Beyond the entrance hall, they could gaze at a globe — seemingly floating in space in front of a starry sky — while a disembodied voice invited them to contemplate their place in the universe. A few exhibits presented introductory concepts, such as an introduction to fossils and an exploration of the way in which past environments can be interpreted. A large window gave a view into the preparation laboratory, while phones allowed access to a narration about the work in progress.

Then the exhibits gradually climbed into a mezzanine, taking Earth's story from the beginning. As the viewer climbed, glimpses down into the great dinosaur hall became possible, and eventually visitors descended (with an opportunity to circle through the palaeoconservatory, with 110 species of plants) into a corner of the great hall devoted to the Mesozoic era. There, forty percent of the exhibit space was given over to dinosaurs and their contemporaries.

"The marvellous exhibits of the RTMP include more skeletons of more kinds of dinosaurs than any other museum on earth," said researcher Peter Dodson. And indeed, "dinosaurs are the meat and potatoes of the Tyrrell Museum of Palaeontology," as freelance exhibit designer Jean André commented. They are indeed what most visitors came to see. More than eight hundred original fossils and thirty fibreglass skeletons and recreations were shown in the museum galleries in 1985. Many of the skeletons were arranged in front of a 900-square-metre (1080-square-yard) (21 feet) mural, 6.5 metres high, enlarged by computer from paintings created by Czech-Canadian artist Vladimir Krb.

The dinosaur skeletons — forty-three by 1998 — were largely fibreglass casts (which are much lighter to mount, and allow the originals to be accessed by scientists when needed). They included material from the Tyrrell's own collection and Canadian material borrowed from or cast by such museums as the American Museum of Natural History, the National Museum of Natural Sciences, and the Royal Ontario Museum. Other exhibits of non-Canadian material came from a number of American museums, often provided in exchange for casts of material in the Tyrrell collection. As well as skeletons, life-size models by Calgary sculptors Brian Cooley and Mary Ann Wilson assisted visitors to envisage the dinosaurs in life, and sections of dinosaur skeletons and bone beds in their rocky matrix showed how dinosaurs occur in the field.

Older animals from the Jurassic period of the United States were the first major dinosaur fauna represented. Dinosaurs and other creatures from three different Cretaceous formations from western Canada followed. An enclosed area with sound effects suggested the submarine environment of the Bearpaw Sea, where marine reptiles such as mosasaurs, plesiosaurs, and turtles could be observed from below.

Visitors made their way back to the entrance through exhibits of Tertiary mammals, and finished with an ice age exhibit and a hint of the origin of people — the only animal species capable of (or interested in) putting together such an exhibit.

Media played an important part of the exhibits. Touchable rocks and fossils provided direct access to the past at many points. Mini-theatres and stand-alone monitors presented short video loops showing fieldwork in progress. Ten pairs of computers were provided for visitors' use, with double seats to encourage people to work together. Half a dozen educational games, and access to a data bank, provided a variety of ways of enhancing the exhibits.

TOO MANY STORIES?

The scale and drama of the exhibits, and their popular subject matter, meant that (as far as published data show) they were generally well received by the public at large. Two museum professionals, John Storer (a vertebrate palaeontologist), and Bob Peart (an evaluation specialist), published commentaries on the exhibits in the early years. Some aspects appealed greatly to the museum professionals, but others received criticism. One reviewer was concerned about the confusing floor plan, and both experts had concerns about the "seemingly endless introduction whose displays treat principles of astronomy, geology, and biology," and the fact that "... little of the introductory material ... makes a clear link between earth history and the dinosaur." Storer commented on the "short and remarkably uninformative narration" that accompanied the globe. Both had overall concerns about the lack of a clear focus, and the attempt to tell too many stories. "The three historical narratives, of earth, of palaeontology, and of dinosaurs, may bear many points of coincidence, but told together they produce confusion." Some basic questions also remained unanswered, such as those about warm-blooded dinosaurs, bird origins, and the extinction controversy.

"Perhaps ..." concluded Peart's evaluation, "when I next visit, I will step through the entrance into that wonderful vestibule and be face to face with *Tyrannosaurus rex.*" Even if this has not yet happened, arrivals at the Tyrrell Museum are now greeted outside by life-size models of dinosaurs leaping towards them along the walls. (Although they give a very odd effect to winter visitors, when the dinosaurs may be liberally decorated with fresh snow, for most of the year they vividly express both the dynamic view most scientists now have of dinosaurs, and the liveliness of the museum within.) Many other changes have taken place, some in response to the criticisms, and others through the evolution of the collections and changing public interest.

TOO BAD THEY'RE ALL DEAD

As originally planned, the Tyrrell Museum has been an active force in palaeontology, both as a research institution and as a public presence. Its staff at opening numbered thirty-one (not including security, maintenance, or seasonal personnel) — an astonishing growth from the two half positions for palaeontology at the Provincial Museum less than two decades before. Six of these thirty-one staff were curators, continuing the multidisciplinary approach started by Currie for research in Dinosaur Provincial Park. Although personnel changes periodically, staff usually has included several vertebrate palaeontologists with different areas of expertise, a paleobotanist, and a stratigrapher. Later directors Emlyn Koster and Bruce Naylor have come from the research side, while a balance of strong public programming has continued through staff with appropriate expertise. More recent research at the Tyrrell Museum will be discussed in later chapters.

Once open, the museum made a determined attempt to become widely connected with the international research community. It offered its facilities as a base for a series of major conferences related to its mandate. In quick succession, the museum hosted conferences on the Systematics of Dinosaurs (1986), the Fourth International Symposium on Mesozoic Terrestrial Ecosystems (1987), the Society of Vertebrate Paleontologists (1988), a planning meeting on Non-marine Cretaceous Correlations (1989), and (jointly with the Provincial Museum) the Society for the Preservation of Natural History Collections (1989). Professional palaeontologists and others from the U.S. have been very enthusiastic about the museum. "In the size and scope of its vertebrate paleontology collection, it beggars every other museum in the world," commented John Horner from Montana. *National Geographic* photographer and dinosaur writer Louis Psihoyos calls it "The crown jewel of paleontological [sic] museums."

Attendance figures continued to astonish all those involved in planning; around half a million people a year make their way to the badlands destination, and the museum received its two millionth visitor in August 1989. Nearby Calgary produces around one-third of all visitors, part of the fifty percent of visitors originating in Alberta. From further afield, around thirty percent of visitors come from the rest of Canada, and twenty percent from the United States, Europe, and the

Pacific Rim. On one of my recent visits I checked the visitors' book and found that the previous month had brought in tourists from twenty countries, as far afield as Australia, Finland, Guatemala, Kenya, Korea, India, Malaysia, and Romania. Visitor responses are generally highly enthusiastic; comments included the serious ("Thank you for the life-long experience given to these children.") and the political ("Absolutely World-class! Thank you, and shame on Klein and budget cuts to a resource like this."). Regular visitors still left their comments ("Fourth time this year — I love it."). Others treated it as entertainment ("Almost better than the real thing." "Fun-Fun-Fun."). And, of course, the inevitable ("It's like *Jurassic Park.* Great!" and "Too bad they're all dead.")

Popular support was formalized when the Friends of the Tyrrell Museum was formed in 1987. The Friends were given a more important role when government policies changed to allow them to collect and manage donations, more recently replaced by an entry fee. As government support dwindled during federal and provincial austerity programs, other revenue-gathering programs were put in place, including fees for programs, taped tour rentals, and donations. By 1990 the Friends had already put more than $1 million into the museum.

In that same year the museum received a promotional boost when — during a visit from Queen Elizabeth II on June 28, 1990 — it was granted permission to add "Royal" to its name — only the third museum in Canada to be so designated.

The locals' overall reactions to the museum varied. Those who had wanted a small addition to their downtown museum were at first bemused — the museum remains open and also serves as a tourist information centre. Some businesses boomed — new hotels and other tourist-related enterprises were among thirty-two new businesses created. However, downtown residents had many reservations, feeling that the museum was too far away to provide all of the expected economic benefits.

ALWAYS THE BRIDESMAID?

"The park is the heart of the dinosaur story and it seems unfortunate that the museum is located so far away," comments Bob Peart on the Tyrrell's site; "Dinosaur Provincial Park remains my wistfully preferred location." And indeed, from the point of view of the fossil resource, Dinosaur Park would have been a much better location. Staff would have been on the spot to research their major field site, and visitors could have

A dinosaur's grave marker.

readily related the park (largely without visible dinosaurs) to the exhibits. Granted, the integrity of the badlands in Dinosaur Park would have been lost by building such a large museum within them — Midland Provincial Park, where the museum was actually built, is a reclaimed coal mine, not an area of great natural beauty. But surely a site on the park boundary — or perhaps even on the famous lookout site — would have provided the best of both worlds.

But there would have been strong practical arguments against Dinosaur Park as a location. It is much farther from Calgary, which provides both the chief cultural base for the staff and a third of the Tyrrell's visitors. Brooks, although now a larger town than Drumheller, is less attractively situated. And — from a strictly mercenary perspective — two major dinosaur sites in southern Alberta offer the opportunity to attract and hold tourists for longer, with associated economic benefits.

To my knowledge, Brooks was never a serious option for the museum. Drumheller had spent years building up its reputation as the dinosaur capital of Alberta, and was the municipality that asked for an infusion of capital. However, the citizens of Brooks, long accustomed to the low profile of their leading attraction, felt that once again they had been bypassed. Dinosaurs were being collected in "their" park, and instead of being shipped off to Ottawa or New York, were now being taken to Drumheller. Only in recent years have the Tyrrell staff been able to develop a strong program working with Brooks-area residents, who now provide important volunteer support and are actively involved in the Tyrrell's programs.

Perhaps partly as a consolation prize, and partly to meet the practical needs of the park and the Tyrrell staff working there, the September 7, 1983 *Brooks Bulletin* announced that a Tyrrell Museum Field Station would be constructed in Dinosaur Park. Construction of the $2.1 million building began in 1986, and the station was opened on May 15, 1987, by Alberta's Minister of Culture and Multiculturalism, nearly two years after the opening of the Tyrrell Museum.

The Field Station has an information centre, and a 262-square-metre exhibit area focusing on the ongoing research in the park. A spectacular centrepiece shows a *Lambeosaurus* being attacked by a pack of four dromaeosaurs. A larger area provides a lab, visible through a window, as in the main Tyrrell building, where visitors can often see fossils being removed from their jackets and prepared. Offices are used by the museum staff working in the area and the parks interpretation program. An orientation theatre shows programming related to the park, and allows groups to attend interpretive programs, which are supplemented by self-guiding trails and guided hikes and bus tours into the "closed area" of the park to see current excavations. The theatre is called the Sternberg-Anderson Theatre, recognizing both Dr. Anderson, who was a prime mover in establishing the park, and Charlie Sternberg, who did so much fieldwork in the area. Of the three great, pioneer Canadian vertebrate palaeontologists who devoted substantial portions of their lives to making the dinosaur resources of western Canada known — Lambe, Parks, and Charlie Sternberg — the latter is the only one to get even this token recognition in the area.

Dinosaur Park has had its resources enhanced, and is even more interesting to visit than before. The Field Station had its 100,000th visitor in May 1989. Sadly, Dinosaur Park is still missed by many who make their way to the Tyrrell Museum at Drumheller.

8

CANADA'S DINOSAURS

"Once, when photographing some badlands, I dropped my camera case down a hill. As I clambered down to retrieve it, you can imagine my surprise when I found it lying directly on a skull of Daspletosaurus, *a smaller cousin of* Tyrannosaurus. *The specimen, which included most of the skeleton, had been entombed for 75 million years, and mine were the first human eyes to see it. Moments like this are exhilarating and deeply moving. Your breath is stopped for a moment when you realize this fossil is a window into the prehistoric world in which this animal fed and rested, had babies, bled and hurt like animals we know today."*

PHILIP CURRIE

BRINGING DINOSAURS TO LIFE

Dale Russell refers to the excitement of each new discovery as "the holy moment." From a background of bone scrap and isolated limb bones, from the damaged specimens field staff dismiss as "road kill," special discoveries stand out. Each skull, unusual bone, complete foot, or crest provides information; a complete skeleton is a special joy, allowing hundreds of isolated bones to be linked to a complete animal. The excitement of such a special find is what sustains palaeontologists in the hard physical work of finding, collecting, and preparing dinosaur fossils, and the equally hard intellectual labour of describing and interpreting them.

For the scientist — and for other dinosaur enthusiasts — a list of names, a shelf full of scientific descriptions, even a gallery full of displayed fossils is only the beginning. Each discovery provides a window into a prehistoric world, and the ultimate excitement comes in peeking through that window — combining previous knowledge with new insight to understand the dinosaurs and the world they lived in.

How many kinds were there? What space and time did they inhabit? Where did they come from and how did they evolve? What did they eat, and how did they obtain their food? How did they behave towards each other, and towards the other life forms they shared their world with? What did the living dinosaurs look like, sound like, act like? How did they reproduce, care for their young, mature, and die? And why did they (apparently) disappear?

HOW MANY DINOSAURS?

Several hundred genera of dinosaurs have been described, but their distribution is uneven. Which country has produced the most dinosaurs? A list published in 1990 included forty-nine genera that had been recorded in Canada. (This does not include ichnogenera, names given to a particular kind of trackway.) At that time, Canada was placed fourth in a "dinosaur league table" behind the U.S. (110 genera), China (96), and Mongolia (50). The closest runners-up included Britain (47) and Argentina (37). Since then, there have been many new discoveries in China particularly, and some in the other countries.

Lists like these don't need to be taken too seriously. A long list of genera certainly may mean lots of dinosaur genera clearly based on thorough study of good skeletons. Equally, it may mean that there are lots of old names based on uncertain material, so that a bit of re-examination would reduce the numbers significantly. However, the league table gives a rough indication of the international importance of Canada's dinosaur resource.

A summary of Canada's dinosaur fauna from the same source included a dozen theropods, a hypsilophodontid, the only known thescelosaurid, eleven hadrosaurs, four boneheads, six armoured dinosaurs, and eleven horned dinosaurs. Today, we would add early Nova Scotia dinosaurs, and a number of others resulting from recent work.

SEA TO SEA . . . TO SEA

Dinosaurs are known to have lived on earth from about 225 m.y. ago to 65 m.y. years ago — an awe-inspiring 160 million years from Late Triassic through the Jurassic period to the end of the Cretaceous period. However, Canada has an incomplete sequence of Mesozoic rocks, and dinosaur fossils are not equally distributed. Many dinosaurs have left no remains; their bodies were scavenged and decayed, their bones scattered on the surface until they crumbled away. Dinosaur remains were deposited in places where sediments were formed on or adjacent to land. Some of these sediments have been exposed and eroded away during the last 65 million years. Others are still deeply buried under later sediments, particularly the glacial drift that the ice age glaciers spread over most of Canada's landscape. There are no doubt thousands of wonderful dinosaur fossils under the rolling prairies of Alberta and Saskatchewan, most of which we will never know about because they will never be exposed to the eager eyes of a collector. Even where dinosaur-rich beds reach the surface, the rocks may be hidden by prairie grassland, forest, or urban sprawl. We have access to fossils only where erosion or excavation exposes rock at the surface.

In the Mesozoic, the landscape that is now Canada did not exist in its present form. During the Triassic, the ancient continent we call Pangaea was beginning to break up. A sedimentary basin opened up through what is now New England and the Maritime provinces, and another further east began what is now the Atlantic Ocean. Through the Jurassic, the northern continent (now North America, Greenland, and Eurasia) remained together, but gradually drifted apart from the united southern continents. In the Cretaceous, this gap widened, but North America was still joined to Asia across the North Pole. Dinosaurs travelled wherever land connections allowed.

The ancient Canadian Shield (widely exposed in Ontario, Quebec, and Nunavut), and other deposits formed around it, make up the core of North America. Rocks formed in the rift valleys through New England and Nova Scotia give us the first view of dinosaurs on the continent in the Late Triassic and Early Jurassic. Through the same period, a large shallow sea spread and retreated across the region of the midwestern states and provinces; in Canada some of its beds are preserved deep under the prairies, and in the Rockies, but contain marine fossils. Jurassic dinosaurs surely walked across the land areas of the western provinces and Ontario, but their remains (including *Stegosaurus* and the giant sauropods) have been preserved only in the U.S. Midwest.

Early Cretaceous seas spread across the Arctic Islands region and down the Pacific coast, with embayments into the northern Alberta area. Sediments are left from the adjacent land areas, including sandstones, shales, and coals. Fossils include many plants, and dinosaur tracks have been preserved in parts of Alberta and British Columbia. By the Late Cretaceous, a sea ran through the modern Great Plains region, at times connecting the Arctic Ocean and the Gulf of Mexico. Rivers running east from the rising Rocky Mountains ran through the lands occupied by dinosaurs, and have preserved some of their remains in the deltaic deposits exposed where modern rivers have cut down to expose them. More recently, dinosaur-bearing Upper Cretaceous deposits have been found in Vancouver Island, Nunavut, and Yukon.

Until recently, dinosaur specialists generally lost interest in rocks younger than the Cretaceous. However, many palaeontologists now believe that one group of direct dinosaur descendants — the birds — has continued to the present, and that we should also consider fossil birds.

AGES OF CANADIAN DINOSAURS

This summary of the dinosaur-bearing beds in Canada shows ages in millions of years before the present. Fm = Formation.

Period	Division	Age	Strata	Key Location
Cretaceous	Late	72–65	Edmonton Group	Drumheller
		79–74	Judith River Group	Dinosaur Park
	Early	146–97	Various fms	Foothills and Rockies
Jurassic	Late	157–146	Mist Mountain Fm	Foothills and Rockies
	Middle	178–157		(No dinosaurs found)
	Early	208–178	McCoy-Brook Fm	Nova Scotia
Triassic	Late	231	Blomidon and Wolfville Fms	Nova Scotia

WHAT KINDS OF DINOSAURS?

So far, we have looked at dinosaurs in the context of their discovery and description. It is time to get a clearer idea of the main groups that have been identified in Canada, and to see how they relate to the dinosaurs known from other parts of the world.

Dinosaurs were regarded as a single group for less than half a century after Richard Owen first named them. In 1887 an astute English palaeontologist, Harry Govier Seeley, recognized that they fell into two groups, which he named Ornithischia ("bird-hipped") and Saurischia ("reptile-hipped"). Although they were assumed to have a common ancestor, it was not known. Both groups were still referred to as dinosaurs, rather in the same way that our vague term "fish" describes several very distinct groups of vertebrates that happen to have fins and live in the water. It was only with the revival of interest in dinosaurs in the 1970s that the order Dinosauria was revived as a formal classification, with Ornithischia and Saurischia as subdivisions. The group now includes recently discovered very primitive forms that are not clearly either ornithischian or saurischian. And although there is not yet complete agreement among specialists, Dinosauria now usually includes birds. (Therefore, there is no good single word for what used to be called dinosaurs — they are often now clumsily referred to as the "non-avian dinosaurs.")

Within the dinosaurs, many groups of related forms have traditionally been placed in families, suborders, and other divisions. When we

Early carnivore from Nova Scotia.

talk about such groups as duckbills and horned dinosaurs, they are the same kind of broad groups that a bird-watcher talks about when discussing ducks or owls.

A bird-watcher can distinguish a species from a group of close relatives — finches, say, or gulls — by looking at its plumage. If handed a bag full of bones from the same species, a bird-watcher might find it quite easy to know he was looking at a finch or a gull, but much harder to be certain what species he was looking at. That, roughly, is the plight of the vertebrate palaeontologist; genera can be distinguished by characteristic bones, such as skulls, but species can be distinguished only by careful examination and measurements. For this reason, we will only discuss genera, and not species.

DINOSAUR BEGINNINGS

The oldest dinosaurs known are from Argentina, and are about 225 m.y. old. In Nova Scotia, as Dale Russell pithily puts it, dinosaurs are "very uncommon and very small." The earliest, from about 220 m.y. ago, are too fragmentary to be fully understood on their own, so we must compare them with early dinosaurs found elsewhere.

One appears to be an ancestral ornithischian. The evidence is an upper jaw, 1.25 centimetres (0.5 inch) long, found by Donald Baird at Burntcoat Head, and a few foot bones. About the same age, tracks at Paddy's Island, near Wolfville, named *Atreipus*, show prints of hands as well as feet, and seem to have been left by a similar small ornithischian dinosaur.

In the same area are three-toed tracks that look as if they were made by a small theropod. Although no bones have been found associated with the tracks, a similar carnivore is well known from New Mexico, where its abundant remains have been named *Coelophysis.* This 2-metre (6 foot) long, razor-toothed predator, balanced on strong hind legs, may have hunted in packs.

Jurassic material found at Wasson Bluff and other sites includes both ornithischian and saurischian dinosaurs. Bird-hipped dinosaurs belonged to a group known as the fabrosaurs, and may have been a genus known as *Lesothosaurus.* Its name comes from the region of Lesotho in South Africa, an indication of the freedom with which small dinosaurs could move between Africa and North America before the Atlantic Ocean opened up. Imagine a dinosaur about the size of a large chicken, with a long tail balancing the neck and head. It probably had a horny beak used for shearing through tough plant material.

Casting a large dinosaur footprint, Peace River Canyon, B.C.

Saurischians include the partial skeletons recently discovered, which belong to the prosauropods — smaller ancestors of the huge, long-necked, long-tailed monsters of later times. It may have been 2.5 metres (7.5 feet) long — a monster in its time. As there is no skull, proper identification is difficult, and several candidates from New England, Arizona, and South Africa have been put forward. Dale Russell favours *Massospondylus,* a genus whose name means "massive vertebrae," first described by Owen from South Africa, but later found in Arizona. One of Russell's books figures the slim but substantial four-legged beast crossing sand dunes, an environment in which skeletons have been found. Others suggest *Ammosaurus*, known from nearby New England.

As described in the opening chapter, prosauropods probably used their long necks — and perhaps stood on their hind legs — to browse trees. We know from the first Nova Scotia skeleton that the prosauropods swallowed pebbles, which would be used to help grind up food material. A mystery is the jaw of a sphenodontid (the group that includes the tuatara, a living fossil from New Zealand) among the stomach contents of this early vegetarian. It was perhaps picked up by mistake with the pebbles, or swallowed for its calcium content. Large tracks (*Otozoum*) may be prosauropod tracks.

The presence of smaller carnivores is indicated by teeth, which may have belonged to *Syntarsus.* Known from Tanzania and Arizona, this genus was larger and more heavily built than *Coelophysis,* and evidence suggests that it ate smaller vertebrates. It may have made the tracks called *Grallator*.

Another track, *Coelurosaurichnus,* has a long claw on its second digit, and may be made by an ancestor of the later "terrible claw" *Deinonychus.* Larger trackways suggest larger carnivores, which could have been the 6-metre (20-foot) *Dilophosaurus*. Its name means "two crests," from the lightly constructed features carried on top of the head. Its jaws were not as strong as those of later carnivores, and it may have scavenged rather than actively pursued the prosauropods and other creatures of the time. Nevertheless, as the largest predator of the time, it perhaps deserved its nickname as "the Terror of the Early Jurassic."

TRACKS IN THE WEST

Many dinosaur track sites from the Upper Jurassic and Lower Cretaceous of western Canada also indicate dinosaurs from both major groups, although attempts to identify track-makers from tracks are not always successful.

Ornithischian dinosaurs are mainly bipeds with three-toed tracks. Some are almost certainly duckbills, such as the common *Amblydactylus* from the Peace River, represented by juveniles as well as adults. Others have been considered to resemble iguanodonts (better known from Europe). Of particular interest are the Peace River quadrupedal tracks that Sternberg called *Tetrapodosaurus,* which may have been made by an ancestral horned dinosaur.

Saurischians from these beds are all carnivores. Tracks of small, lightly built coelurosaur-like theropods have been named *Columbosauripus*. Larger theropod tracks have been called *Irenisauripus*, and have been found in a number of sites in the Crowsnest area and the Peace River Canyon.

One of the most spectacular sites is on the Narraway River near Dinosaur Ridge, west of Grande Cache, Alberta. There, two hundred footprints have been recorded in at least eight trackways. Most of these are small theropods. One large biped with feet half a metre (18 inches) long provides us with a prehistoric moment, for: "This individual came to a stop, turned sharply to the right, paused and then set off in a direction perpendicular to the original route."

TRACKS AND TRACK-MAKERS

Well-preserved tracks offer many clues to the track-maker. The task is easier if beds of the same age and region contain abundant dinosaur fauna, so that one can compare tracks to known animals. The width of the trackway can indicate if the animal had feet immediately below and supporting the body (as in dinosaurs) or whether the legs were widely splayed (as in amphibians and many primitive reptiles).

The size of the tracks suggests the size of the animal, while the shape of the tracks and the number of toes are obvious clues to the track-maker. More subtle details such as the relative thickness of the toes, the angle between them, the presence or absence of claws and hooves, and indications of hard pads can also tell much of the anatomy of the feet.

The pattern of a trackway can indicate whether the animal puts two or four feet on the ground, and show if the forefeet are used regularly or only occasionally. However, forefeet may be much smaller and make a light impression, so they may be missed. Sometimes the regular gait of a dinosaur may place its hind feet on top of forefoot impressions, and they will be revealed only when a track changes direction. After much careful comparison, there will still be tracks that are hard to identify.

Other aspects of dinosaur tracks are used to reveal the position of the animal in life — for instance, the presence or absence of a dragging tail. Its habitat may also be suggested; some footprints may have been made under water, others in drier ground. Formulae have been devised to estimate the speed with which a given animal was moving. Of particular interest are instances where the trackway suggests aspects of behaviour that cannot otherwise be guessed at, such as travel in groups or pursuit of other species.

UPPER CRETACEOUS

	Judith River	Horseshoe Canyon	Scollard
Genera Described	39	20	14
Ornithischia			
Armoured (Ankylosauria)			
Ankylosaurs	*Euoplocephalus*	*Euoplocephalus*	*Ankylosaurus*
Nodosaurs	*Panoplosaurus*	*Edmontonia*	–
Boneheads (Pachycephalosauria)			
Pachycephalosaurs	*Pachycephalosaurus*	*Stegoceras*	*Stegoceras*
Horned (Ceratopsia)			
Protoceratopsidae	–	–	*Leptoceratops*
Centrosaurinae	*Centrosaurus*	–	–
Ceratopsiae	*Styracosaurus*	*Pachyrhinosaurus*	–
Chasmosaurinae	*Chasmosaurus*	–	*Triceratops*
Ornithopods (Ornithopoda)			
Hypsilophontidae	–	–	–
Thescelosauridae	*Thescelosaurus*	–	*Thescelosaurus*
Duckbills			
(Hadrosaurinae)	*Gryposaurus*	*Edmontosaurus*	–
(Lambeosaurinae)	*Parasaurolophus*	*Hypacrosaurus*	–
Saurischia			
"Segnosaurs" (Therizinosauroidea)	*Erlikosaurus*	–	–
Theropoda			
Elmisauridae	*Elmisaurus*	–	–
(Caenagnathidae)	*Chirostenotes*	*Chirostenotes*	–
Dromaeosauridae	*Dromaeosaurus*	*Saurornitholestes*	–
Ornithomimosauria	*Ornithomimus*	*Ornithomimus*	*Struthiomimus*
Tyrannosauridae	*Albertosaurus*	*Albertosaurus*	*Tyrannosaurus*
Troodontidae	*Troodon*	*Troodon*	*Troodon*
(Birds)	*Apatornis*	(unidentified)	(unidentified)

UPPER CRETACEOUS DINOSAURS

Once the Canadian Upper Cretaceous is reached, almost all major groups of dinosaurs are represented in the rich fauna. Dinosaur classification has been in an ongoing state of flux, particularly in recent years as specialists use a new approach known as cladistics for studying relationships, which are also illuminated by a flood of new material from Asia.

Within the Ornithischia, the groups include both quadrupeds and bipeds. The quadrupeds are grouped into armoured (Ankylosauria), boneheads (Pachycephalosauria), and horned (Ceratopsia). The bipeds (some of which seemed to have occasionally used their forefeet) include duckbills (Hadrosauridae) and the Thescelosauridae.

The Canadian saurischians were all carnivores, of various sizes. The early simple grouping into large and small carnivores has now been replaced by more sophisticated classification, and the Canadian carnivores are now classified in five or six groups. The segnosaurs are a primitive side-shoot from the main evolutionary line. The caenagnathids were toothless and at first regarded as birds. The large carnivores are now placed in the tyrannosaurids. The troodonts are small, delicate carnivores, while the ornithomimids are the tall, slim bird mimics that looked like ostriches without (presumably) feathers. Lastly, the dromaeosaurs were small, active carnivores with a prominent claw on the hind feet.

ARMOUR, SPIKES, AND TAIL CLUBS (ANKYLOSAURS)

Like turtles and armadillos, ankylosaurs found safety by developing body armour. This has its price; the heavier the armour, the shorter their legs must be (to keep the centre of gravity low). Weight and short legs combine to reduce mobility, so that they must have been able to find an adequate supply of food close to the ground and within a reasonable distance. Ankylosaurs were the largest fully armoured animals that have ever lived, yet they may have been able to move with the efficiency of a modern rhinoceros.

The armour is formed by bony scutes that grew within the skin, and that are often found as loose fossils. Between the large scutes are many small, bony nodules, which may have extended over the belly and legs. Their skeletons usually are found upside down, the carcass settling in this position when rolled by rivers. The fronts of their jaws were without teeth, but they probably cropped plants within a couple of metres of the ground using horny beaks. Tongue bones suggested that they had mobile tongues, and gastroliths (stomach stones) helped some of them to process hard vegetation.

Ankylosaurs arose in the Early Jurassic and were most diversified at the end of the Cretaceous. They show two lines of development. The nodosaurs had narrower skulls, and some had striking spines along the shoulders (perhaps for intimidation or even for aggressive responses to predator attacks). The ankylosaurs (in the narrow sense) had wide skulls, and a club at the end of the tail that was so solid it sometimes became fossilized when the rest of the carcass decayed.

The nodosaur *Panoplosaurus* from the Judith River Group was about 5.5 metres (18 feet) long, and its plates had only small spikes. *Edmontonia* from the Edmonton group was even larger, growing to 7 metres (23 feet). It has substantial spikes along the flanks, and the shoulder spike is divided into two. This was perhaps used in combat between males, in the same way that deer lock antlers and wrestle for supremacy. One example with damaged jaw and tail suggests injuries in life.

In the ankylosaur line, the Judith River genera include *Euoplocephalus,* whose name has been translated as "well-protected head." It deserves its name, for it even has curved disks of bone protecting the eyelids. It was 5 metres (17 feet) long, and (like other ankylosaurids) breathed through complex sinuses that may have provided a

Styracosaurus *out for a stroll.*

strong sense of smell, been used to warm the air, or provided a resonating chamber for the animal's calls. Its tail was built of ten vertebrae, fused and strengthened by supporting bony prongs. The club was composed of four fused vertebrae and would have been lighter in life than the mineralized masses found today. The jury is out on the mobility of tail and club; some feel its movement was quite restricted, while other restorations show it being swung from side to side like a medieval mace.

Its presumed descendant *Ankylosaurus* was the last (and one of the largest) of its group at 7.5 metres (25 feet). It was rather less armoured than *Euoplocephalus* — perhaps its size helped to protect it from aggression. Swinton has called it "the most ponderous animated citadel the world has ever seen."

HITTING IT OFF — ANCIENT BONEHEADS

The boneheads are generally uncommon fossils, nowhere known from a complete skeleton. The largest and commonest is *Pachycephalosaurus,* which is known from the Judith River and possibly the Edmonton groups. With a body size of 4.5 metres (15 feet), this bipedal dinosaur was probably an active runner, with its tail (supported by ossified tendons) held out stiffly. Its most remarkable feature was a skull 0.6 metres (2 feet) long. Serrated teeth once regarded as *Troodon* were probably used for cutting leaves and twigs. The bones over the forehead and crown were fused into a single bone mass 23 centimetres (9 inches) or more thick, which was surrounded with blunt spines. Although this profile suits our image of a high, intellectual forehead, the brain within this solid mass was quite small, though large olfactory lobes suggest it had a good sense of smell.

There is broad agreement that this solid skull was probably used for competition between males. The most common suggestion is that the males sorted out their "pecking order" by running at each other and butting heads, as in modern Bighorn sheep. This view is supported by occasional specimens marked by scars. Among those to study this problem was Hans-Dieter Sues (now of the Royal Ontario Museum), working with Peter Galton in Connecticut. In the 1970s Sues modelled the collisions. Stress lines produced in the plastic models mimic the patterns of bone columns in the skulls, which help to pass the destructive forces around the skull in the same way that a football player's helmet protects the head. An alternative possibility that the heavy skulls were used to butt each other in the ribs has also been argued. The smaller *Stegoceras* lived in the Edmonton times and was about 2 metres (7 feet) long.

HORNS AND FRILLS — THE CERATOPSIANS

Western Canada has around ten genera of horned dinosaurs. They arose in the Lower Cretaceous, and reached their full flowering in the Upper Cretaceous. All are rather rhino-like hoofed quadrupeds, that ate plants with the aid of a horny beak and strong teeth. Most have an assortment of horns and a huge bony frill extending from the back of the skull. The skeleton shows holes in the frills of some kinds, but these were probably filled with muscle and covered with skin, so would not have been apparent in a living animal. Various explanations have been offered for the frill — defence and offence, the support of large jaw muscles, or dispersal of excess heat.

However — influenced by evidence that shows these were herd animals — some scientists seek parallels for horned dinosaurs in modern mammals. The antlers of deer, the herding of antelopes, and the protective rings of musk ox all suggest possible behaviour patterns. The elaborate frills may have been for purposes of sexual display — to terrify rival males and impress females. This view is particularly popular with dinosaur artists, as it gives them licence to add striking colour patterns to the impressive shapes. One implication of this theory is that the females may be less elaborately decorated than the males. If so, previously described genera with simpler ornamentation may not be separate kinds, but females of other "genera." For this reason, the relationships — or even the identity — of some genera of horned dinosaurs are in doubt.

If we stay with more conventional taxonomy, ancestral forms are small and hornless. An example is *Leptoceratops*, under 2 metres (6 feet) long, which seems close to the ancestry of the line. Oddly, it survived until almost the end of the Cretaceous, when it could have walked under the belly of its larger relative *Triceratops.*

Two different evolutionary lines — short-frilled and long-frilled — can be distinguished. The first can be represented by *Centrosaurus* ("sharp-point reptile"), known from a few skeletons and the extensive work on the Dinosaur Park bone bed. It grew to around 6 metres (19 feet), with a skull 1 metre (3 feet) long. It had a large nose horn up to 47 centimetres (18 inches) long, and small brow horns. Two more horns hung down over the holes in the short frill. (For a few years after 1989, *Centrosaurus* had a short-lived career under the name *Eucentrosaurus.* It was renamed when it was discovered that the earlier name had previously been used for a lizard. However, further research showed that the original lizard usage was not scientifically valid, so the dinosaur name did not have to change.)

Close relatives include the spectacular *Styracosaurus* ("spiked reptile"), with its frill edged by huge spines. Also related is *Pachyrhinosaurus* ("thick-nosed reptile"), which occurs in the Edmonton group. Its massive head apparently had no horns; however, new material from Pipestone Creek near Grande Prairie suggests that some individuals — perhaps of a different species — bore three straight horns on the midline of the frill.

The long-frilled ceratopsians include *Chasmosaurus,* also from the Judith River rocks. Up to 5.8 metres (17 feet) long, its impressive frill has two huge holes, the source of its name — familiar to us from the use of the word "chasm" for a large opening in the ground. It had a small horn on its nose and two on its forehead.

Despite its size, *Chasmosaurus* is still described as "modest-sized" beside its successors, which include the Edmonton group *Triceratops* ("three-horned face"). At 9 metres (30 feet) this monster must have weighed around 5.4 tonnes. Its horns were up to 90 centimetres (3 feet) long — "as long as hockey sticks," says Tyrrell Assistant Director Monty

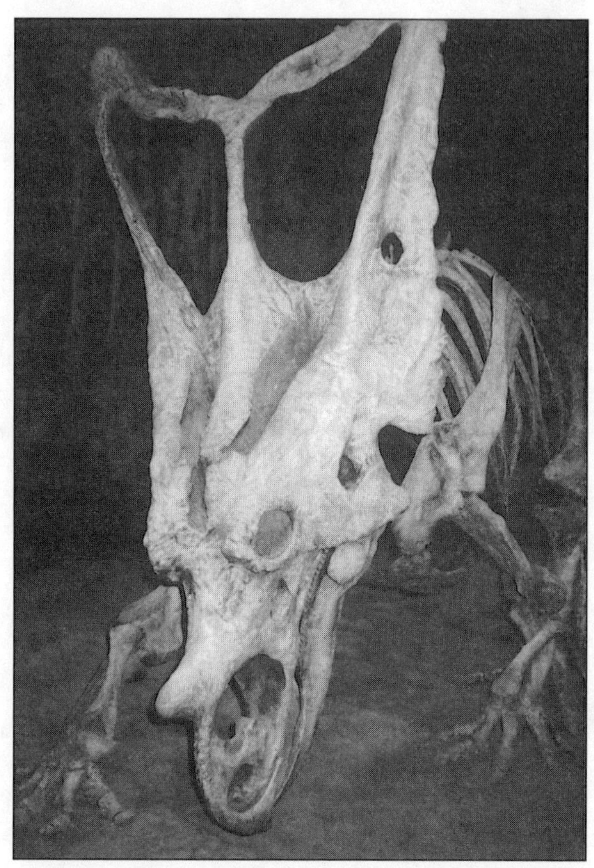

Chasmosaurus shows off a long frill.

Reid, using a very Canadian image. These were long enough to lock and wrestle like deer, and fights are suggested by many damaged bones. *Triceratops* had strong jaws and shearing teeth, which may have been used for cutting fibrous material like cycad and palm fronds.

NESTS AND CRESTS (DUCKBILLS AND RELATIVES)

Although Canada's Upper Cretaceous is known for great numbers (both individuals and kinds) of such groups as horned dinosaurs and duckbills, the richness of the fauna is also reflected in other groups that are represented by only one or two genera. One of these is the thescelosaurids, a group of small plant eaters. They first appeared in the Middle Jurassic, but are better known from the Cretaceous. *Thescelosaurus* ("marvellous reptile") was 3.5 metres (12 feet) long. It had a moderately long neck and a long tail. Generally lightly built, its heavy hind legs suggest a runner. It is not clear if it was a biped or a quadruped, and may have walked both ways. It may have lived in or near streams, where its remains are sometimes abundant. *Thescelosaurus* is one of most primitive of its group, which is ironic since it occurs among the latest dinosaurs. Teeth are found in Dinosaur Park, and skeletal material in the Edmonton series.

More familiar relatives are the duck-billed dinosaurs, or hadrosaurs. A dozen genera are currently recognized in western Canada. The most obvious distinction between them is that some are crested and some are flat-headed. At one time it was suggested that these represented different sexes, the males (as so often among animals) having the impressive crests. It then became customary to distinguish the flatheads and cresteds as different evolutionary lines, but it seems the relationships are more subtle than that. Recent workers recognize the hadrosaurine subfamily (with flat heads, or crests that are made of solid bone). The lambeosaurines have hollow crests.

Duckbills were vegetarians, with batteries of many teeth — as many as two thousand — combined into a single unit able to grind up tough vegetation. Mummies of *Edmontosaurus* found by the Sternbergs in Wyoming had eaten conifer twigs and needles — not easy chewing. Earlier views saw them as amphibious, but most associated plants seem to have grown on land. The duckbills' long hind legs perhaps allowed them to reach up into trees to seek food, but they probably walked on all fours at least part of the time. Many of them seem to have travelled in herds. Their tail vertebrae were linked together by a complex of tendons, which probably supported them horizontally.

Crestless duckbills include the big-nosed *Gryposaurus,* known from the Judith River Formation, and *Edmontosaurus,* named after the formation in which it was found (which, in turn, was named for Alberta's capital city). *Edmontosaurus* is known from a number of skeletons, some of which have extensive skin impressions. At 15 metres (48 feet) long, it was probably the largest and perhaps one of most abundant of the group and seems to have occupied damp lowlands close to the seashore. It may have had brightly coloured balloon-like flaps of skin on the face that inflated when the animal called.

Maiasaura, the first dinosaur found with extensive nests, eggs, and even embryos in Montana, also belongs to this group, and has now been found in Alberta.

The hollow-crested dinosaurs include *Lambeosaurus* and *Parasaurolophus,* which both occur in the Judith River beds, and *Hypacrosaurus,* owner of the nests found in Devil's Coulee. *Lambeosaurus* is the largest of the group, up to 15 metres, (46.5 feet) long, and the heads of different species are crowned with a variety of crests. Most striking is the rare *Parasaurolophus,* whose length was around 9 metres (28 feet), and which may have weighed 3.5 tonnes. Its crest extended up to 1 metre (3 feet) behind the head.

Lambe had realized that the air passages led through the crest, and looped back into the skull. CAT scans of a very complete skull found recently in New Mexico show as many as six paired tubes. Many possible explanations have been advanced. While duckbills were presumed to be aquatic, a snorkel was one obvious explanation, but no hint of nostrils has been found at the tip. An air reservoir is unlikely, but space for a rich supply of smell receptors is possible. Other suggestions include a support for muscles activating an elephant-like trunk, space for salt glands, deflectors of foliage like a cassowary's crest, and a temperature regulation device. Certainly such a prominent feature was probably useful in visual recognition — perhaps enhanced by bright colours? However, the favourite function is that of the trombone the bent tube so closely resembles — to produce deep resonant calls. One American researcher, David Weishampel, has gone so far as to construct his own horn (a parasaurolophonium?) on which he blows melancholy notes. "We can imagine," says English palaeontologist Michael Benton, "the Late Cretaceous plains of Canada and Mongolia reverberating to deep growls and blaring squawks as the hadrosaurs went about their business."

THE MISSING SAUROPODS

Where's *Diplodocus*? Although most people are aware that dinosaurs come in many kinds, the ultimate dinosaur seems to be a sauropod. Long-necked and long-tailed, the public expects them from any area that claims to have dinosaurs. Why doesn't Canada have any?

There are two great groups of Saurischians — the reptile-hipped dinosaurs — which superficially appear very different. One group, the bipedal carnivores, are widespread both geographically and stratigraphically. The other group is the prosauropod-sauropod line. The prosauropods were widely distributed, and occur in the Nova Scotia Jurassic, as we have already seen. Their descendants, the giant sauropods (which include the largest dinosaurs — and the longest of any animals — known), are more restricted in their distribution.

These great dinosaurs were discovered in the American Midwest, and the names of "*Brontosaurus*," *Diplodocus*, and the like soon became household words. In more recent decades, bigger and bigger animals were discovered, and dinosaur aficionados have followed the progression of *Brachiosaurus* to *Supersaurus* to *Seismosaurus* to *Argentinosaurus* with awe.

The sauropods appeared in the middle of the Jurassic period, a time when Canada has left no dinosaur-bearing deposits. They flourished in the Upper Jurassic and on into the early Cretaceous in midwestern states. Although less well known from the Upper Cretaceous, sauropods of this age have been found in parts of the U.S., the southern continents, Europe, and Mongolia. Some sauropods have been reported as close to Canada as Montana; *Apatosaurus* and *Camarasaurus* from the Upper Jurassic, and *Alamosaurus* from the Upper Cretaceous.

Why do they not — so far as the evidence shows — occur in Canada? Certainly they have been expected by Canadian palaeontologists. Langston commented on "Non-marine sediments of Late Jurassic and early Cretaceous age . . . where the discovery of large sauropod . . . dinosaurs would not be surprising." As Phil Currie systematically explored many trackway sites from Upper Jurassic and Lower Cretaceous, he anticipated the possibility of coming across the huge, rounded tracks that would say "sauropods were here." Now he comments: "There would have been no physical restrictions preventing their access to western Canada throughout the Cretaceous. The complete absence of sauropod tracks and bones is a good indication that sauropods did not penetrate into Canada for ecological reasons . . ." A detailed analysis of the environments in which dinosaur footprints are preserved shows that coastal plains with coals and other evidence of vegetation are associated with hadrosaurs and their relatives. Sauropod footprints typically are associated with lake deposits and drying coastal plains. Other research suggests they principally fed on conifers, less common in the beds studied in Canada.

Canadian bird-watchers are accustomed to having to seek the more exotic birds — parrots, ostriches, and penguins — in other climates. It seems that the dinosaur-watcher must do the same if an acquaintance with sauropods is desired. But perhaps, somewhere, in the right environment, remains of the major missing group of dinosaurs are waiting to be found in Canada.

STRANGE "SEGNOSAURS"

In Mongolian religion, Erlik was the King of the Dead, and an odd reptile from Mongolia discovered in 1979 was named after him. Its 5-metre (17-foot) body had the long neck and tail of a sauropod, yet claws on its feet were like the carnivores. It had a toothless beak, and small, pointed teeth further back in the jaw. Surprisingly, *Erlikosaurus* solved the problem of a skull fragment found in Dinosaur Park that could not be related to any other kind of dinosaur. There, in 1982, Phil Currie found a claw that resembled that of *Erlikosaurus.* A bigger puzzle is the relationships of this odd dinosaur (and a few of its relatives), which was formerly placed by different researchers with the sauropod group, but now seems to be a specialized carnivorous dinosaur.

Carnivorous dinosaurs fall into two major groups. The primitive ceratosaurs included the early carnivores of Nova Scotia, but the Late Cretaceous carnivores of the west all belong to a later branch that arose in the Middle Jurassic, the Tetanurae ("stiff tails"), all of which had tails supported by interlocking projections of the vertebrae. Within this group several distinct types have been recognized, as different as modern carnivores — bears, wolves, cats, and so on. The dinosauria carnivores include the small elmisaurids and troodonts, the large carnivores such as *Albertosaurus* and *Tyrannosaurus,* the tall, slim bird mimics, such as *Ornithomimus,* and the sickle-clawed dromaeosaurs.

THE HAND BONE IS CONNECTED TO THE FOOT BONE

Bones of a slender "hand" described as *Chirostenotes,* and a foot called *Macrophalangia* were connected when a more complete skeleton was found in 1979 that had the hand of one and the foot of the other. Although still not complete, it was clear that the dinosaur belonged to a group known as the elmisaurids ("foot reptiles"), found in Mongolia. It was 2 metres (7 feet) long, and had a long, slender clawed finger that might have been used for prying into cavities in trees or rocks; the curved teeth found in a lower jaw suggest it was carnivorous. It is now known as *Chirostenotes.* Dale Russell notes that it has a large pelvic opening, and "may have given birth to living young."

HEAVY-DUTY CARNIVORES — TYRANNOSAURS

Everyone's favourite monsters, the large carnivores are as familiar as any dinosaurs to the public. Yet they are not well represented in the rocks. *T. rex,* the biggest carnivore, is only known from twenty-three reasonably complete North American specimens, of which two come from Alberta. Other western Canadian large carnivores include *Albertosaurus* and *Daspletosaurus.* An open question is whether the large carnivores were active hunters or scavengers. Lambe considered that their teeth were not worn, and envisaged them as very passive, lying on the ground until hunger forced them to go in search of decaying carcasses to scavenge. More recently, it has been suggested that they would lie in wait in cover, then catch their prey in a short, vicious rush — up to 40 km/h (25 mph) — as do many mammalian carnivores today. Probably they did both, as opportunity offered. Lighter juveniles may have pursued smaller prey more actively than the adults. The short forelimbs — not long enough to reach to the mouth — were a puzzle for many years until it was suggested that they would anchor the body of an animal rising from the ground until its long back legs could get underneath and take the weight.

Albertosaurus is the best-known carnivore — half a dozen new skeletons have been found since the Tyrrell Museum opened. It also ranged very widely; a carnivore that is probably *Albertosaurus* has been recorded in Alaska and a Canadian Arctic island. Specimens range from quarter-size juveniles to full-grown animals. At 10 metres (31 feet) an adult was only a little smaller than *Tyrannosaurus,* but its limbs were longer in proportion, so it was probably a faster runner, and Dale Russell has suggested that it might have specialized in hunting the fast-moving duckbills. Its clawed feet could have delivered a devastating kick, and its long teeth, serrated like a steak knife, allowed it to bite an unprotected neck and then tear off chunks of meat from its victim. The teeth often broke on bones, and are found loose at sites where prey have been consumed. They were replaced in waves, so often present an irregular appearance in the jaws. By contrast, *Daspletosaurus* was more heavily built, and may have specialized in hunting ceratopsians.

Alberta's specimens of *Tyrannosaurus* came from Huxley in the Red Deer Valley and the Crowsnest Pass. The Huxley specimen was found by Charlie Sternberg in 1946, but the bones were in a hard ironstone that made a formidable obstacle to collection. In 1960 Wann Langston found bits of the skull below the cliff. In 1981 the Tyrrell staff attacked the cliff with jackhammers and pneumatic drills, and obtained most of the skeleton, but the skull end of the specimen was under 27

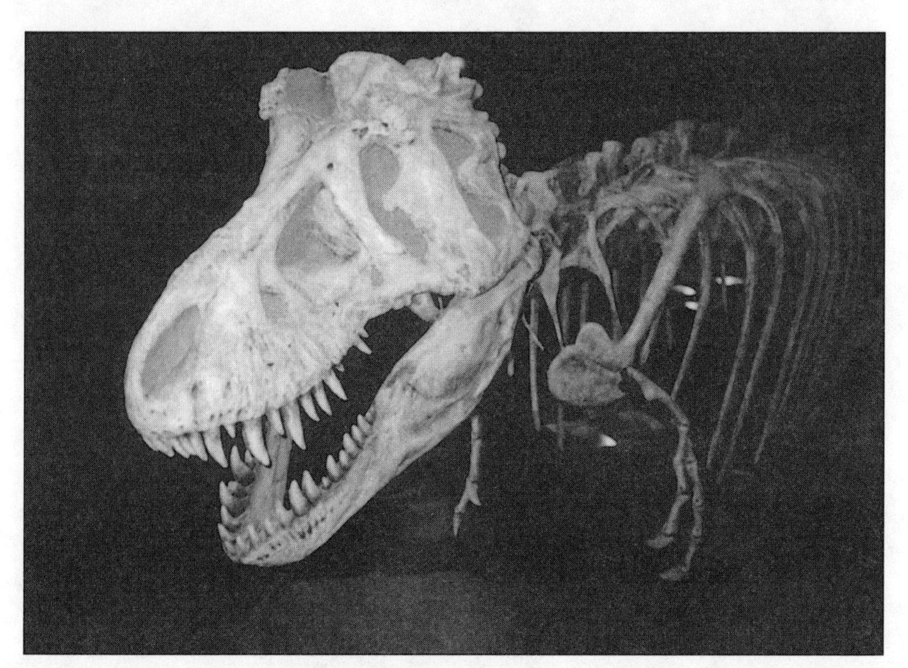

Albertosaurus, *a formidable predator.*

metres (90 feet) of rock. In the same year, three high school students found another *T. rex* in the Crowsnest Pass. Its bones were stained black by manganese, so they stood out from the reddish sandstone matrix. Nicknamed "Black Beauty," it also was collected by the Tyrrell, and went on tour with the Ex Terra exhibition. Both of these were large animals, at about 12 metres (40 feet) long.

The dagger-shaped teeth alone were up to 16 centimetres (6 inches) long, able to saw meat and break bones. Phil Currie has studied the 2-metre (six-foot) skulls of tyrannosaurs with John Horner in Montana. Some of the bones are several inches thick ("heavy duty," says Currie), but there are also big holes, which reduce the overall weight and provide access for nerves, blood vessels, and muscles. CAT scans are being used to allow a close look at the complex structure of the interior. Its brain is as long as a human brain — one of the largest of any animal. Damaged skulls suggest they sometimes attacked each other.

A NAME FOR A CARNIVORE

The carnivorous dinosaurs found in western Canada presented problems for those who tried to name them. Palaeontologists first try to compare new material with fossils that have already been named. If the fossil belongs to a kind already described, naming it is a simple matter. If it clearly belongs to a kind not already described (and that could be anywhere in the world, in any one of several languages accepted by science), then the discoverer has the right to describe it and give it his or her choice of name. Problems come when it is not clear whether it is new or not — imagine trying to compare a handful of teeth with a species someone has named from a pelvis and limb bones.

A very sketchy look at the tangled history of the names used for the large carnivorous dinosaurs of western Canada will give an idea of the difficulties met by palaeontologists working in the area.

Laelaps was the first dinosaur named by Cope himself, from a few teeth, a piece of jaw, some neck vertebrae, and leg and toe bones. It was found in Upper Cretaceous rocks of the New Jersey coast, and is thus roughly the same age as the Canadian specimens. Its name is a delightful joke, for it is that of a dog in Greek mythology that was turned to stone. Cope saw it as a "totally new gigantic carnivorous Dinosaurian, probably of Buckland's genus *Megalosaurus*" and his contemporary sketch shows it standing in a kangaroo-like stance.

Unfortunately for Cope, his rival Marsh delighted in pointing out in 1877 that the name had been used already in the animal kingdom (for a spider!), and proposed the new name *Dryptosaurus.* (Even then this was probably gross discourtesy; nowadays a palaeontologist might point out another's error but would wait for the original describer to come up with a new name). Although Cope refused to use Marsh's name, it became the "official name" for this dinosaur.

In 1892, Cope described the skulls found by Tyrrell and Weston, assigning them to a species, *Laelaps incrassatus*, which he had already named from some teeth found in Montana.

In 1905, Osborn was a little vague about the geography of Canada (he refers to *Dryptosaurus* finds as being from British Columbia), but he recognized that the inadequately defined *Laelaps* is not a suitable name. He renamed Tyrrell's skull as the type of a new genus and species, *Albertosaurus sarcophagus*. The first name means "Alberta reptile," and is a reference to the newly named province. The second name is Greek for "flesh-eating," commonly used for a tomb because it appears to consume the body.

To complicate matters further, what seemed to be a related dinosaur was described by Lambe in 1914 as *Gorgosaurus,* and for several decades the term was used in preference. In 1970 Dale Russell reviewed the large carnivores, decided these were the same genus, and that *Albertosaurus* had priority. Currie is now resurrecting the name *Gorgosaurus* for the albertosaurs of Dinosaur Park, which he recognizes as distinct from the Drumheller region *Albertosaurus.*

SMALL DINOSAURS WITH BIG BRAINS (TROODONTS)

In 1968, Dale Russell and colleagues spent six weeks of a hot summer in a largely unsuccessful search for small carnivores in Dinosaur Park. In frustration, he told park naturalist Hope Johnson what he was looking for. Johnson remembered having seen just such material in a shoebox belonging to a local fossil collector, Irene Vanderloh. She showed her specimens to Russell, but was doubtful that she could find the site again after six years had elapsed. Her memory was good; Gilles Danis, technician with the party, found part of the skull of *Stenonychosaurus.* This is typical of the few chance finds that have gradually pieced together the story of the troodonts.

Troodon ("wounding tooth") has one of the most chequered careers of any Canadian dinosaur. Leidy named it in 1856, on the basis of a few teeth from Montana, and considered it a lizard. It was half a century before scientists began to suggest it was a dinosaur. In 1924 Gilmore found what he considered similar teeth in jaws of one of the bonehead dinosaurs, and for a while took over the name for a dinosaur we now know is quite unrelated. Gradually, a few jaws with teeth were found, and it became clear that *Troodon* was a carnivore, but there was little research material for further study.

For years afterwards, the small carnivores were largely neglected, until the 1970s, when both Currie and Dale Russell began to systematically collect new material and review it. Fortunate discoveries like the one mentioned above have totally changed our understanding of this group. Another was the one made by John Horner at the Tyrrell Museum site (discussed in the previous chapter) that showed that *Troodon* and *Stenonychosaurus* (and two other genera) were the same animal. Currie's study of the material humorously acknowledges Horner's find "in my own back yard at the Tyrrell Museum, after I stepped over it. And it must have taken a special courage for him to refind the site two years later once it became evident that the specimen was going to disprove one of his pet theories."

We now know that the troodonts were little carnivorous dinosaurs, about 1.7 metres (6 feet) long. Dale Russell calls them "Cretaceous coyotes." They were bipedal, and had long necks and tails. The long, slender hind legs ended in feet with big claws, and it would have been able to grasp prey with its long, clawed fingers. Large eyes facing forward suggest activity at low light levels and the ability to judge distance. Most striking are the other features of the skull, for this little dinosaur had the biggest brain of any dinosaur (in relation to its size), and it had evolved complex sinus systems on the base and sides of the brain case that are much like those of living birds. One of these connects the two ears under the brain case, perhaps an adaptation to allow location of sounds.

Troodon's large, forward-looking eyes and its potential for speed and agility suggest it went after some fast-moving prey. It would have been well adapted to hunt the small mammals of the time. Dale Russell suggests that "the hands were adapted to grasp a victim while it was being eviscerated by the talons on the inside of the feet."

OSTRICHES WITHOUT FEATHERS (ORNITHOMIMIDS)

Imagine a living ostrich without feathers. Add a long bony tail, and you have a good idea of this group of carnivores. The first to be found was named *Ornithomimus* ("bird mimic") by Marsh; a later genus, Osborn's *Struthiomimus,* would have given us a better name for the group, as "ostrich mimic" is more precise. They arose in the Middle Cretaceous and flourished in the Late Cretaceous, so many of the important specimens have been found in western Canada.

Ornithomimus was around 3.5 metres (12 feet) long, and would have stood taller than a human. With its light build and long legs, it is usually viewed as the fastest dinosaur, with speeds of 40 km/h (25 mph). Its small but strong head had lost most of its teeth, but its mouth was probably strengthened with a horny beak or sharp scales to help it grip its food.

There have been many suggestions about ornithomimid diet, including eggs, insects, plants, and an omnivorous lifestyle. A careful look at the body elements used for food gathering and processing can help resolve the problem. It has been suggested that the clawed hands may have been used for digging ants from their nests, but a 1985 study by Canadians Elizabeth Nicholls and Anthony Russell have suggested that the fingers may have been opposable, and thus more suitable for grasping. This accords with Osborn's suggestion that *Ornithomimus* was a browsing herbivore, pulling branches down with its hands to

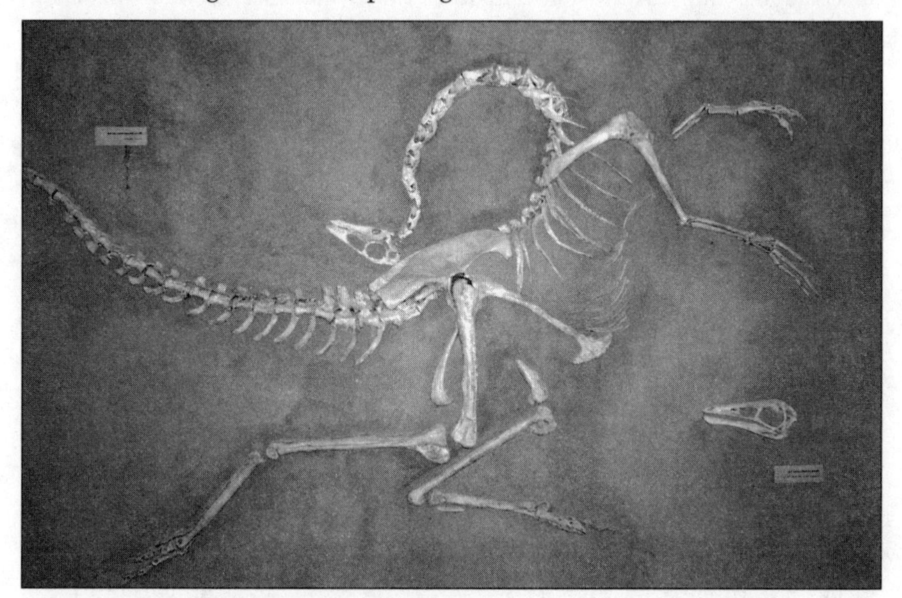

Ornithominid *skeleton distorted in death.*

pick the tastier leaves. On the other hand, there is no evidence of gastroliths, used to grind plant material.

Its forward-looking eyes would see in stereo, a help in catching moving prey. The weak jaw muscles suggest soft food, such as fruit or insect larvae. The skull may have been flexible, as in modern birds and snakes — a suggestion supporting an egg diet. An unusual suggestion was made by Brown, who noted the frequency of *Ornithomimus* in shore deposits, and suggested a diet of crabs and shrimps. This would suggest a comparison closer to a heron than an ostrich — but the toes do not seem long and flexible enough for this lifestyle. There is in fact some evidence in support of almost every kind of prey suggested — which may indeed be the truth. It is easy to imagine *Ornithomimus* as an active seeker of food in a wide variety of habitats, picking up whatever it could catch or find in season. All the time, it would be seeking alertly for danger, ready at any moment to take to its heels.

GRIM REAPERS OF THE CRETACEOUS (DROMAEOSAURS)

Barnum Brown discovered a partial skeleton of the first *Dromaeosaurus* (its name means "swift-running reptile"), but it has been understood only in more recent years, when more complete relatives have been found in the U.S. and Mongolia. The group arose in the Middle Cretaceous, and was most abundant in the Upper Cretaceous, but has so far been found only on the northern continents.

Dromaeosaurus grew to about 1.8 metres (6 feet). Although Dale Russell suggests that "the animal fed probably on carrion left behind on tyrannosaur kills," it seems to be particularly designed for attack. A large head, with a mouthful of sharp notched teeth, was supported by grasping hands. The secret weapon was the middle toe, around 8 centimetres (3 inches) long, with a sickle-shaped claw. Just as cats can retract their claws, so could the dromaeosaurs bend theirs upwards out of the way, to protect them from wear when running over rough ground. It is easy to envisage *Dromaeosaurus* hunting, until "when needed, [the claw] could be flexed back into lethal position, and with the powerful kicking motion of the rest of the leg, this razor-bladed foot could slash its way into the belly of some hapless herbivore, disembowelling the animal in one rapid stroke." Another dinosaur researcher, Greg Paul, offers a different scenario: "They probably ran alongside their prey and leaped onto their backs. Using their long, strong, big-clawed fingers to hold on, they could then wound the prey, leaping off before the latter could roll over and crush them."

Either scenario is unpleasant for the victim, but when the little dinosaurs had "... an extraordinary degree of agility, balance, and leaping ability" and probably hunted in packs, it is easy to see that the one thought of a potential prey animal would be not to get caught by what Monty Reid calls the "grim reaper of the Cretaceous."

FEATHERED DINOSAURS

It is widely suggested that small carnivores such as troodonts and dromaeosaurs may have had feathers, and are so closely related to birds that birds, too, may be regarded as dinosaurs. The considerable contributions of Canadian researchers to this question came about substantially through the Ex Terra project (see Chapter 11). But in view of the wide (though by no means universal) agreement of many palaeontologists that birds are in fact dinosaurs, it seems appropriate to mention at least the Cretaceous Canadian birds here.

Currie's discovery of bird tracks from the Peace River shows that they were present in Canada by the Early Cretaceous. Although flying animals are rarely fossilized, once we reach the Upper Cretaceous, the intensive search for small bones has yielded a few bird fossils from a number of sites in several provinces and as far north as the Arctic.

A traditional expression for something non-existent is "as rare as hen's teeth." Although hens don't have teeth, a number of the oldest known birds did. One of the first birds known from fossils was *Hesperornis,* in appearance a loon-like swimming bird. Richard Fox has reported it from Alberta, and it is also known from Manitoba. A nesting colony was suspected by Dale Russell in the vicinity of the Anderson River, in the Northwest Territories, because so many juveniles were discovered. Further discoveries have been made on Bylot Island (off Baffin Island) and Eglinton Island.

Another early discovery, *Ichthyornis,* was a superficially tern-like bird that also had teeth. It seems to have been a fish eater, and ranged widely over Cretaceous seas. Through work by Richard Fox and Larry Martin, its remains are now known from Alberta ("the oldest skeletal fossil of a bird from the Western Hemisphere") and Manitoba. Another Cretaceous bird, *Apatornis,* is less well known, and it is not certain if it had teeth or not. Dinosaur Park has yielded an "eagle-sized flying bird" close to *Apatornis.*

Saskatchewan has yielded a rich bird deposit along the Carrot River, and Royal Saskatchewan Museum staff Tim Tokaryk, John Storer, and Stephen Cumbaa have described more than one hundred fossils representing seventeen different types of birds. This, says fossil bird expert Alan Feduccia, is "the oldest and most diverse avifauna from the North American Cretaceous." This includes *Ichthyornis,* a primitive relation of *Hesperornis,* and several other kinds.

These discoveries, made in such a short time, suggest that more bird remains will be found in the Canadian Cretaceous, even though they are but a tantalizing glimpse of the birds that might actually have flown over the other dinosaurs. Another chapter would have to be added to this book to pull together the ongoing story of the avian dinosaurs (birds) through the 65 million years since the rest of the dinosaurs disappeared. The giant, flightless *Diatryma* in the Northwest Territories hints at how fascinating such a chapter might be. And, of course, the story of the modern birds (recent dinosaurs) of Canada is already the matter of many books.

However, it certainly adds a new dimension to the dinosaurs to think that some of them have flown out of the Cretaceous period and are with us today. As I write, the rufous hummingbirds come to the feeder at my window. If, as I believe, the birds-as-dinosaurs palaeontologists are right, my hummingbird has genetic material in common with the dromaeosaurs.

9

DINOSAUR LIFE AND DEATH

"It is probable that when this formation was deposited the country had been sufficiently elevated to drain off the marshes, and that the draining of the waters was the chief cause of the extinction of the dinosaurs."

BARNUM BROWN

DINOSAUR WORLDS

From almost the earliest days of dinosaur discovery, scientists theorized about their mysterious disappearance. Barnum Brown is discussing the Paskapoo formation, deposited immediately after the dinosaurs became extinct. His simplistic interpretation suggests that if the food of the plant-eating species (then regarded as swamp dwellers) became scarce, it would explain their extinction. Like most palaeontologists, Brown was not just interested in dinosaurs. Anything collectable was added to his boxes: scutes of crocodiles, pieces of fossil wood, teeth of small mammals. All kinds of evidence went to round out the picture of the world that the dinosaurs lived in, the world of which they were then regarded as the primary inhabitants.

During the century and a half that has elapsed since the dinosaurs were first recognized, a vast amount of evidence has been accumulated, not only about the dinosaurs, but also about the changing world in which they lived. It is certainly easier to gain a picture of the landscapes and plants that formed the stage on which they walked. And we now know the dinosaurs are just some of the players — though they include the largest and most spectacular — that strutted those ancient stages. We also realize that dinosaur behaviour and ecology must have been vastly more complex than was at first thought, yet it has become much easier to see them as living, breathing animals — herding and breeding, eating and being eaten. And (though answers sometimes seem overwhelmed with contradictory data) the last great question, that of their disappearance, continues to fascinate scientist and layperson alike.

DELTAS AND OXBOWS

In 1983, the Calgary Zoo created a new prehistoric park. Existing dinosaur models were renovated, and some splendid new ones were created. "The designers of the new park," commented Banff writer and curator Jon Whyte, "suffering Disney spells, set their huge, wonderful, fiber glass dinosaur replicas in a place that resembles the Red Deer badlands, not that region as it looked when the last saurian bellowed on the shore of the Bearpaw Sea." Sadly, this choice of setting perpetuates the erroneous idea of many badlands visitors, that the dinosaurs lived in the landscape that they see today. Although in theory there may have been some kind of barren rock exposures at the time the dinosaurs lived, the

Stream channel deposits in Dinosaur Park.

badlands we see today are postglacial, at the most a few thousand years old. Abundant evidence shows that the places the dinosaurs inhabited were quite different.

The early landscapes inhabited by the east coast Jurassic dinosaurs included sandy stream deposits, and were close to active volcanoes. Other sites of deposition include a lava-dammed lake (in which muddy limestone was deposited) and active sand dunes.

In Alberta, volcanic rocks in today's Crowsnest Pass and abundant volcanic ash embedded in bentonite-rich badland deposits tell us of active volcanoes in the mountains rising to the west through the Cretaceous — the beginnings of today's Rocky Mountains. Rivers running east from these crossed the uplands and plains, eventually running through estuaries and deltas into the warm, shallow arm of the sea that ran north-south across what is now eastern Alberta and Saskatchewan.

These rivers constantly changed their courses, as do unchannelled rivers today. Had time-lapse photography been possible, we might have had a picture of rivers writhing slowly across the landscape, cutting sideways here, depositing sandbanks there, abandoning arms that became oxbow lakes and then slowly filled with sediment to become swamp and then dry land. Low-lying ground was periodically flooded by overflowing rivers, and the sea level itself changed from time to time, flooding shallow ground near the shore with salt water — and leaving such deposits as the oyster bed along Drumheller's Dinosaur Trail. From the abundance of fine particles in mudstones, Dale Russell suggests that the rivers were generally slow moving — half a kilometre (a quarter of a mile) an hour. Deposits in bars and deltas show that larger streams were around 6 metres (15 feet) deep. Levees along river shores are shown by roots of trees, and other deposits indicate filled-in oxbow lakes and widespread floodplains. In higher deposits, swamps often gathered deep layers of peat, which have become the Drumheller area coal seams. The general picture was similar to the modern Mississippi delta.

WARM POLES AND STORM-SWEPT PLAINS

When we look at the tree rings in a cut stump, we are using a simple tool to understand palaeoclimatology. In the northern latitudes of Canada, the alternate dense and less-dense wood speaks of seasonal change. Narrower tree rings suggest poorer growth, perhaps from a dry growing season. A scar grown over by new wood can result from forest fire damage — itself an indicator of hot, dry weather. If the tree has been cut down in our lifetime, we can supplement its information by weather records going back decades, even a century or so. Very ancient trees can give us indications back a few hundred years — well beyond the human instrumental record of weather, but not beyond records of such extremes as floods, droughts, and remarkable cold spells. In some parts of the world, such as the American southwest, tree rings can give indications of conditions up to a few thousand years ago, helping to understand the archaeological record of people before written documentation.

Similar plant indicators can be found for the ages of the dinosaurs. With other data, such as geological deposits characteristic of dry or wet climates, and a knowledge of the position of the land in relation to the poles, we have at least indications of the climate the dinosaurs enjoyed. In general, the earth was warmer than it is today, a result of higher carbon dioxide levels increasing the "greenhouse effect."

The Jurassic of what is now Canada's east coast had a warm and wet climate, with a seasonal dry period of perhaps two or three months (during which crayfish could survive deep in sand). During the wet period, flash floods overflowed the rivers; drying sediments preserved tracks, and then even drier weather caused the mud to crack.

Later, in the Cretaceous, the pole was situated over the present Bering Strait, so that today's Alaska and the Yukon would have been (as they are today) in the polar region. There was a long polar night, but a humid and frost-free climate allowed trees — and dinosaurs — to flourish, at least in the summer. Warmer latitudes would be reached by travel to western Canada, where there was still a strong seasonality — shown by growth rings in both trees and the bones of crocodile-like champsosaurs. "It was probably a Georgia or Florida climate," says Dennis Braman, palynologist at the Royal Tyrrell Museum. The mean temperature was perhaps 15 to 25 degrees Celsius and rainfall was abundant, but there were also periods dry enough to allow the woods to burn, leaving scraps of charcoal in sand deposits. Further from the pole in the Appalachian area, there was a mean temperature of 20 to 30 degrees Celsius and a moderate rainfall distributed through the year.

Even more-localized weather events might have left their traces in the rocks. Dave Eberth, sedimentologist at the Tyrrell, has been studying the origins of the *Centrosaurus* and other bone beds of the Red Deer River. "They can be hundreds of metres in lateral extent," he explains. "They are all monogeneric, with 95% a single genus of ceratopsian. The take home message is that they were clearly herding, with the size of groups larger than any documented before, with herds in the high thousands/low ten thousand range, suffering from single killing events."

He compares the gently sloping landscape whose evidence is preserved in Dinosaur Park to modern hurricane-torn lands such as Bangladesh and the eastern U.S. "There are two types of flooding events in a lowlands coastal setting." One is when a river overflows, the other is a coastal flood. "A storm surge can block the drainage of the coastal plain — there is nowhere for the water to go." Oxygen isotope data from the Bearpaw Sea suggests it was just warm enough to generate monsoons, which happen when a large land mass is adjacent to stable, warm, oceanic waters. "Mudstones that host the ceratopsian bonebeds have a lot of plant debris that typically gives them a reddish-brown colour." This suggests a type of flood associated with high winds and "blow-down" from the forest canopy. This in turn suggests that "these flooding events are occurring in conjunction with 'windy' storms — most likely hurricanes. This in turn is compatible with the notion of seasonal coastal floods, storm surges and drowning of the coastal plain for significant lengths of time."

If this interpretation is correct, the many ceratopsian bone beds in Dinosaur Park record such major storms. But this does not mean that the weather is likely to have been violently stormy most of the time. "Even if major storms were recorded once every 100 to 1000 years, they would appear quite common in a rock record that records 2 million years in 80 metres of rock." Eberth points out that vulnerable areas today have more frequent violent storms than the park record indicates.

AMBER AND POLLEN

All food chains begin with plants, and large numbers of dinosaurs indicate the presence of lots of vegetation. The Jurassic of Nova Scotia is illuminated by deposits of similar age in the adjacent U.S., where lake sediments preserve remains of conifers and cycads, ferns and horsetails.

Western Canada presents a variety of evidence for the plants that surrounded and sustained the later dinosaurs, from 60 centimetre (2-foot)-wide and 17-metre (52-foot)-long tree trunks, beds of coal, globules of amber, and tiny grains of pollen. The structure of the wood and the presence of cones shows that the largest trees are redwoods — relatives of those that still grow in California. These probably grew on higher ground to the west, and their broken trunks were carried into the area where fossils are found. Amber — fossilized resin — came from damaged trunks and branches of these and other conifers. More local plants are found in the form of broken stems and leaves, but these are insufficient to give more than indications of the variety of plants that existed.

Fortunately, many trees and other wind-fertilized plants shed enormous amounts of pollen — sometimes today the ground may be yellow with it in spring. The hard shells of pollen grains are not only usually characteristic in shape, but extremely resistant to decay. The first pollen from Dinosaur Park was found in the fossilized excrement of a crocodile, but the abundance of such remains in other deposits suggested that a systematic search would be productive.

The National Museum of Canada added a palaeobotanist, David Jarzen, to its staff, who had the palynology lab in full-scale operation by December 1974. Samples were taken from the insides of fossil turtles already in the museum collection, and also from outcrops. The rock samples were crushed, and then dissolved in hydrofluoric acid — one of the most powerful known. Chemical and physical sorting of the residue makes it possible to concentrate the pollen, which is mounted on slides and studied under the microscope. As the types of plant found in Canada today are generally very different from those of the Cretaceous period, Jarzen travelled the world to build a reference library of more than sixteen thousand pollen samples, which could be used for comparison with Canadian material.

By 1982, eighty-five different kinds of plants had been identified from the pollen and spores of the Dinosaur Park area. The most abundant were flowering plants, the next most frequent were spores of ferns, mosses, and liverworts, while the smallest fraction were coniferous trees. Only three could not be identified. Subsequent work has greatly extended our knowledge of the plant life. Dennis Braman estimates that there could be as many as five hundred species of terrestrial plants now known from the Alberta Upper Cretaceous. Lots of them are unlike anything now living. "Dinosaur Park has lots of evidence of angiosperms [flowering plants]," says Braman, "but the wood is gymnosperm [conifers and relations]. There could have been big trees along the banks, like the balsam poplars along the Red Deer."

Many kinds of plants would be familiar today — at least to gardeners. Quiet water was bright with the blooms of water lilies, and its margins were fringed with familiar cattails. Dawn redwoods (a favourite for office landscaping) lined higher river banks. Six-foot horsetails flourished along stream banks, perhaps among patches of *Sphagnum* moss, and sycamores and breadfruit were among the trees growing on higher ground. "There was no grass," Braman reminds us. Ground cover was ferns and club mosses of familiar types, sometimes decorated with arums and lilies. Epiphytic plants (which grow directly on other plants) included relatives of the mistletoe. Less-familiar larger plants included tree ferns, cycads, evergreen China Fir and *Torreya*, and broad-leaved *Katsura*.

Higher in the sequence, the Edmonton group deposits show abundance of swamp cypresses as well as the other tree species. Broad-leaved trees included ash, swamp hickory, and swamp oak.

TINY MAMMALS AND GIANT PTEROSAURS

The dinosaurs were not alone, and many kinds of evidence tell us about other animal species. If amber tells us little about the plants it comes from, it can be revealing about the small animal life of the past. The sticky surface of the original resin was a fatal trap to small creatures walking the tree trunks or the ground under the trees. By careful study of these golden drops under the microscope, familiar types of invertebrates can be shown to have lived in dinosaur times. Larger animals provide more substantial evidence in the form of their scales, bones, and tracks.

The earliest Canadian fauna, in the Nova Scotia Triassic, was a world in which dinosaurs existed, but did not dominate. Instead, the dominant creatures were large reptiles of groups largely unknown to the non-palaeontologist. Long-necked marine reptiles up to 6.5 metres (22 feet), crocodile-like "gatorlizards," phytosaurs, armoured aetosaurs, and cumbersome predatory rauisuchids with teeth 6 centimetres (2.5 inches) long, ruled the roost. More modern groups included small, agile crocodiles; the earliest pterosaurs; and the few dinosaurs.

Later, in the early Jurassic of Nova Scotia, still little is known of insects, but some of the small vertebrates were surely insectivorous. Fish lived in the lakes, long-legged crocodiles were active near and in the water, and sphenodonts (lizards related to New Zealand's living fossil tuatara) and "paramammal" trithelodonts prowled the land along with the early "uncommon and very small" dinosaurs of that time.

Amber shows us that the Cretaceous forest floor was occupied by ants, beetles, and termites, and air was sometimes filled with the tiny biting flies we know as "no-see-ums," which doubtless bothered dinosaurs in the past just as they attack dinosaur hunters now. Web-spinning spiders caught and ate some of these insects.

Many kinds of fish included gars, sturgeon up to two metres (six feet) long, bowfins, rays, and sawfish, indicating both fresh and brackish water. Bones of marine plesiosaurs show that these entered freshwater. Salamanders and frogs lived in the lakes and streams. Along the lake and river margins turtles up to 1 metre (3 feet) long sunned themselves on logs, while crocodiles and alligators — up to 2.5 metres (8 feet) long — no doubt pretended to be logs until they could get close enough to their intended prey. Similar reptiles from the now-extinct group of champsosaurs seem to have had similar habits.

Living entirely on land were lizards, up to 1.2 metres (4 feet) long, but snakes seem to have been uncommon in Dinosaur Park times. Bits of the fragile bones of flying reptiles show that the skies were not deserted. These include the largest of all, the 15-metre (50-foot) wingspan *Quetzalcoatlus* — as big as a small plane.

Although fragmentary remains of mammals had been found quite early in the century, little work was done on these until the new techniques of screening and washing came into use. "The history of early mammals," wrote University of Alberta professor Richard Fox, "has represented my central research interest for twenty years. . . . My own fieldwork since 1966 has revealed an unexpectedly rich succession of fossils documenting the final stages of the history of early mammals, from about twenty m.y. before the extinction of the dinosaurs to about 10 m.y. thereafter." This important work has shown the presence of a variety of primitive groups such as the extinct multituberculates (the largest of which was perhaps the size of a modern house cat) and opossum-like marsupials, and their replacement by more modern mammals resembling insectivores. Most mammals of the time were small, and perhaps nocturnal. Fox notes that their "small size and shy habits were probably essential adaptations for survival in a world terrorized by reptilian predators."

BLOOD: HOT, COLD, OR JUST WARM?

Within this changing physical and biological environment lived the dinosaurs. But how did they live? Were they cold and sluggish, incapable of vigorous movement, or were they much more bird- and mammal-like, showing vigorous activity? The "hot-blooded dinosaur" controversy has been one of the most vigorous of recent years, and there is still no general agreement on the solution. The story of Canadian dinosaurs would be incomplete if we did not review some of the evidence derived from Canadian sources, and see where Canadian palaeontologists stand on the issue.

Earlier views of dinosaurs by Cope and Huxley showed them as lively creatures. J.W. Dawson, in his *The Story of the Earth and Man* presents a surprisingly modern view as he discusses "a type of a group of biped bird-like lizards, the most terrible and formidable of rapacious animals that the earth has ever seen. Some of these creatures, in their short deep jaws and heads, resembled the great carnivorous mammels [sic] of modern times, while all the structure of their limbs had a strange and grotesque resemblance to the birds. Nearly all naturalists regard them as reptiles; but in their circulation and their general habit of body they must have approached to the mammalia, and their general habit of body recalls that of the kangaroos. . . . Had we seen the eagle-clawed Laelaps rushing on his prey; throwing his huge bulk perhaps thirty feet through the air, and crushing to earth under his gigantic talons some feebler Hadrosaur, we should have shudderingly preferred the companionship of modern wolves and tigers to that of those savage and gigantic monsters of the Mesozoic."

Yet by the middle of the twentieth century, a much more static view prevailed. Colbert, in his 1962 book *Dinosaurs: Their Discovery and Their World,* suggested that ". . . . we can picture dinosaurs as spending much of their time very quietly, either lying or sitting motionless on the ground, or moving about with slow and deliberate motions. . . ." This picture is very much like — and about as exciting as — a crocodile snoozing beside a zoo pond an hour after feeding time. Loris Russell's pioneering dissenting view — published only three years after Colbert's book — was discussed in Chapter 7.

The terms cold- and hot-blooded that are in popular use are much more simplistic than the terms preferred by scientists. Modern reptiles are regarded as "cold-blooded" because individuals tame enough to be approached typically feel cool to our touch. However, their temperature varies with that of the external environment, so that they are sluggish when it is cold and active when it is hot. Scientists prefer the term ectothermic (dependent on heat from external sources). As typical mammals, we can tell if we are sick by seeing how far our temperature varies from the normal. As with birds, we mammals maintain a uniform temperature whatever the external circumstances, and are considered "hot-blooded" or, more precisely, endothermic (maintaining heat internally). Since dinosaurs are extremely varied, in theory the group could include examples with either physiology, and of a variety of intermediate or different kinds. It has also been recognized that the size of particularly large animals creates a sort of endothermy based on the slower rate of warming and cooling.

SMALL, AGILE THEROPODS

While it is possible (if difficult) to take the temperature of a crocodile while it is warming up, extinction has made it impossible to do this with a non-avian dinosaur. Our evidence must therefore be indirect, and based on various parallels with our knowledge of modern animals. Directly relevant evidence used to interpret dinosaur physiology comes from bone structures, predator-prey ratios, and evidence of insulating skin coverings. Less direct support comes from showing that they had patterns of distribution, behaviour, and reproduction comparable to modern endotherms, such as Arctic survival, herding, nesting, and caring for young.

The bone questions have largely been argued in the U.S. and Europe, though some Canadian material has been used. Thus R.E.H. Reid in Belfast has found zonation in the bone of a small theropod from Dinosaur Park, and suggests that the evidence points to "the animal being more primitive physiologically than typical dinosaurs, and possibly to its reverting to an ectothermic status in its late years." Canada's Dinosaur Park has been intensely discussed for the light it may cast on predator-prey ratios. Canada's (and Alaska's) High Arctic dinosaur records are of significance in considering survival in extreme conditions, and important behavioural indications come from Canadian footprint evidence and nesting sites.

Anyone who has kept snakes knows that they do not need to be fed as often as pet mammals or birds. Without the need to maintain a constant temperature, reptiles have an advantage when food is scarce; they become cold and sluggish for a while, and can fast. Robert Bakker, while looking for evidence that dinosaurs might be hot-blooded, studied predator-prey ratios in different animal communities. The nearest modern equivalent to a dinosaur-dominated food chain may be the Indonesian island of Komodo, where the 3.5-metre (12-foot) Komodo dragon eats its own weight in prey every two months. By contrast, lions eat their own weight in food in about eight days — ten times as much consumption as reptiles. One implication of this is that the same sized population of prey species can support more cold-blooded than warm-blooded predators. This means that the ratio of predators to prey may indicate the physiology of the predators.

Bakker looked for well-documented communities in the past, and found evidence suggesting that pre-Mesozoic communities had far more predators that Tertiary, mammal-dominated examples. He felt that dinosaurs in the rich and well-studied fauna of Upper Cretaceous

Dinosaur Provincial Park showed a much more mammal-like proportion. "In Alberta," said Bakker, "the ratio of predator to prey averaged 4% . . . tyrannosaurs were probably as warm-blooded as the sabertoothed cats and wolf-bears that took over the top predator role many millions of years later."

However, as Dale Russell and Pierre Béland (and others) have pointed out, there are difficulties in interpreting the data. The relative rarity and great interest in carnivorous dinosaurs has created a collecting bias in favour of them that may exaggerate their abundance. Even if the sample is without bias, recent studies in Africa show that in drier climates predators tend to concentrate at watering places, so that they may be overrepresented in the riverine park deposits.

Robert Bakker has argued that Arctic finds of dinosaurs at latitudes where modern reptiles do not live support the idea of warm-blooded dinosaurs. The Yukon site appears to have been at a latitude of around 80°N. However, plant and other evidence suggests that climates of the Cretaceous Arctic seem to have been significantly warmer than those of today, reducing the force of Bakker's argument.

In recent years, scientists have suggested that dinosaur physiology was in various ways intermediate between modern endotherms and ectotherms. Some kinds may have been more endothermic than others — perhaps even the same individual animals may have been different as their life cycle progressed. Adult dinosaurs may have had an intermediate form of physiology, with characteristics of both. Certainly secondary evidence from bone beds shows evidence of huge herds. Tracks provide valuable indicators of behaviour, such as an exciting sequence in the Peace River Canyon that shows a small group of hadrosaurs moving together, and (on being approached by a single carnivore) all changing direction at once in response to the apparent threat.

Recent viewpoints of Canadian palaeontologists are supportive of at least some endothermy. Dale Russell feels that endothermy was evolving in dinosaurs, until by the Late Cretaceous, "Many had brains that were larger in proportion to their bodies than in modern reptiles. These attributes imply that metabolic rates and activity levels had increased to become transitional to those typical of modern warm-blooded birds and mammals."

"Most palaeontologists now believe small agile theropods were probably warm-blooded," says Phil Currie, "but that most other dinosaurs were not." And in the context of small raptors ". . . the stereotype has been supplanted by agile, warm-blooded, intelligent forms more in keeping with our image of birds."

NESTS AND EGGS

A complex family life like that of birds provides strong (if indirect) evidence of endothermy. Some living reptiles — crocodiles, for instance — approach this, making nests communally, guarding the nests, and even carrying newly hatched young to a safe place. But the young can feed themselves, and from then on they are more or less on their own. How can we tell what sort of family life dinosaurs had? For a long time, the puzzle was to find any evidence at all of dinosaur reproduction.

In 1955, Charlie Sternberg described the skull of a very young hadrosaur, found near the Steveville ferry upstream from Dinosaur Park, and compared the abundant nests and eggs in upland deposits in Mongolia with the then-scarce remains of young dinosaurs in Alberta. At the 1963 conference he expanded on this question: "The lack of young is another thing. I have a feeling that the eggs were deposited somewhere above the delta and only the more or less mature dinosaurs came down to the delta. Otherwise we would have found more. We had one little juvenile specimen amongst those I collected and all the rest are half grown to mature. Now there must be some reason."

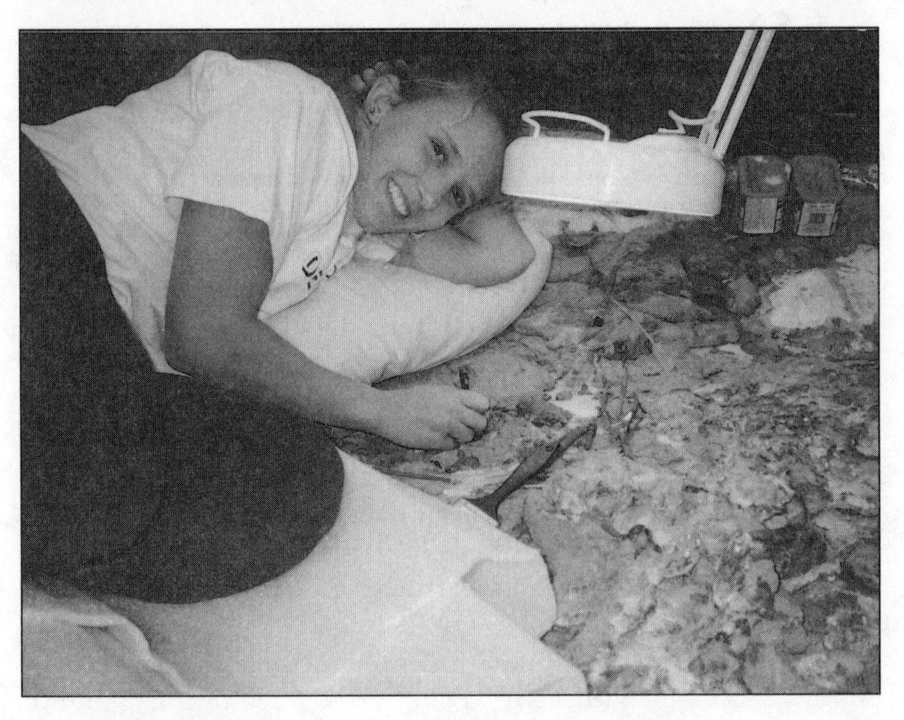

Wendy Sloboda and eggs.

Model of a Troodon.

Sternberg's view, that dinosaurs must have laid their eggs in upland areas, which were usually eroding rather than gathering sediments, was the best answer anyone could come up with. Ironically, abundant eggshells had been found in 1916, just 10 kilometres (6 miles) south of the Canadian border at Landslide Butte, in Montana. The finder was Charles Gilmore, of the Smithsonian — but he hadn't recognized what he had found. Since dinosaur eggshells were not at that time generally known, he assumed the fossil fragments were clams, and their true identity was not discovered until 1986, when John Horner checked the site.

Horner was in a good position to recognize eggs, as he and colleagues had been studying another Montana site called Egg Mountain. Sternberg would have been happy, for the site was in the upland area. Egg Mountain produced many nests and eggs of a duckbill later called *Maiasaura* — the "good mother dinosaur" — and of the little predator *Troodon*. Gilmore's misunderstood eggshells at Landslide Butte proved to belong to another duckbill, *Hypacrosaurus*.

The evidence suggested nesting colonies, with nests clustered together, separated by about the distance a sitting adult can bite — a pattern found in colonial ground-nesting birds today. But the bones of the young dinosaurs showed an interesting difference. Embryonic *Troodon* had well-developed limb bones, suggesting the young were mobile as soon as they were hatched. By contrast, hatchling maiasaurs found in the nest had poorly ossified bones. It seems that young *Troodon* left the nest as soon as they were hatched, like grouse and other ground-nesting birds today. However, the young maiasaurs, Horner argued, could only have survived if they were parented like a baby robin, with an adult looking after them and bringing food to the nest. More than twenty years of work has produced many more nests in Montana, but that *Maiasaura* nest with hatchlings has not been duplicated. However, abundant remains of young inside the colony suggest that they stayed around the nesting site, so the implication of parental care remains.

Troodon and *Hypacrosaurus* have long been known from Alberta, and *Maiasaura* has now been found there. When in 1987 nests were found in Alberta's Devil's Coulee, they proved to be from *Hypacrosaurus.* These were the same age as those in Dinosaur Park, though from a different environment. Like other *Hypacrosaurus* sites found in Montana, they were lowland sites. And here colonies of hypacrosaurs gathered, nesting together, hatching their eggs, and looking after their young until they were ready to leave the nest.

DINOSAUR EGGS IN DEVIL'S COULEE

In May 1987, high school student Wendy Sloboda was searching the Milk River ridge near her home in Warner, south of Lethbridge in southern Alberta, when she found abundant fragments of dinosaur eggshells. North America's major nest finds began only in 1978, when Jack Horner discovered nest and eggs in nearby Montana. Some of these beds are continuous across the international border into Alberta, so it seemed only a matter of time until they were found in Canada.

Tim Schowalter had shown Phil Currie fragments of eggs along the Bow River in the same year, and an association had been noticed by University of Alberta researchers between eggshell fragments and certain kinds of shellfish. Eggshells are calcareous, and normally dissolved in acidic environments in which many fossils were deposited. These shells seemed to provide a buffer for the acid, allowing eggshells to be preserved. With this information, small fragments of shell were found in a number of places in southern Alberta. However, these shells seemed to have been washed from the land into the environments occupied by the fresh-water shells, so these were not places where complete eggs and nests might be found.

By 1986 Horner and his colleague Bob Makela had found many eggs in Montana sites, including one just across the border that had a different chemistry. In that year, a palaeontological impact study was under way for a proposed dam west of the town of Milk River. Staff from the Tyrrell Museum had flown over the site in a helicopter, and when they landed the helicopter pilot asked to see a piece of fossil bone. Looking for something to show him, the team picked up a toe bone from a baby dinosaur within minutes. Plans were made to revisit the area, but had to be put off until the following year. Meanwhile, the impact study directed by Len Hills of the University of Calgary had Sloboda working over the ground in detail. She mailed what she thought might be egg fragments to Hills in a Sucrets box, and he immediately phoned Currie.

By June 1 Sloboda was showing her site to a Tyrrell reconnaissance crew. Egg fragments were abundant in several sites, along with bones and teeth of dinosaurs and other kinds of reptiles, amphibians, and mammals. Sloboda was so excited by the finds that she gave up a paid summer position at the University of Calgary to volunteer with the Tyrrell crew. "Her sharp eyes, quick wit, and enthusiasm never let us down," says Currie.

Back at Sloboda's site later in June, Currie "had the good fortune to discover Alberta's first nest of dinosaur eggs!" Two more nests were found later the same day, and technician Gerhard Maier found "miniature bones of duckbilled dinosaurs strewn all over a hillside." More nests, eggshell, and baby hadrosaur bones were found the next day. The crew was due to leave at noon to return to Drumheller, but technician Kevin Aulenback was late for the rendezvous, and arrived "running over a hill with a bag of baby dinosaur bones . . . talking so fast that it was difficult to follow what he was saying." "The next few minutes," remembers Currie, "were unquestionably the most exciting I have ever experienced as a professional palaeontologist. In a hillside covered with the remains of broken eggs and baby dinosaurs were a number of eggs sectioned by erosion that obviously contained the remains of dinosaur fetuses."

Currie phoned his co-workers, and swore them to secrecy. "We were on top of the world," but: "The following week has to rank as one of the worst in my professional career." The landowner was at first uncooperative, and the story ended up in the *Calgary Herald* and on the wire services. "The first call came through that morning at 6:30 a.m., and calls from media around the world never let up until after midnight. During the morning they averaged one every three minutes. For the next week, I spent most of my time answering telephone calls from radio, television, newspaper, and magazine reporters."

Sloboda, too, was in the middle of a media frenzy. "Reporters were phoning us, following us, flying over our house — Kevin [Aulenback] and I spent a month and a half going everywhere except where the eggs were."

The province moved quickly, and in September the Minister of Culture and Multiculturalism was able to announce the province's purchase of the site, and its designation as a Provincial Historic Site. Now it can be visited from the Interpretation Centre in Warner, run by Wendy Sloboda, who is now on the Tyrrell Museum staff.

A BANG OR A WHIMPER?

There's one thing everyone knows about dinosaurs. "This comet hit the earth, and wiped them all out. Right?"

This question seems to be thrown at every dinosaur specialist on every talk show, and no matter how cautious the reply, only one possible answer seems to remain firmly in the public mind. Some dinosaur specialists feel that 160 million years of living dinosaurs are more deserving of their attention than a possibly unanswerable question about their disappearance, while others join in the fray with enthusiasm. But many dinosaur specialists feel we are not yet sure of the questions, and there is as yet little agreement on the possible answers.

First, perhaps not all dinosaurs are extinct. If birds are their direct descendants, the big question is about the extinction of what are now clumsily referred to as "non-avian dinosaurs" — and why some of the most vulnerable dinosaurs made it into the Tertiary. Second, species, genera, and larger groups of dinosaurs have become extinct on an ongoing basis through the long history of the group (as growing evidence shows that many lines of descent came to an end at different times in the Mesozoic), so why question only the relatively sudden extinctions of larger numbers of species? And if we are to talk of large extinction events, we have to remember that they have occurred at least twice in dinosaur history — at the end of the Triassic, and at the end of the Cretaceous. These are among more than a dozen concentrated extinction events that have occurred at other times in earth history, before and since the dinosaurs lived — could all of these events be connected in some way? And since these major events have been responsible for the disappearance of many living things other than dinosaurs, how can we find an explanation that will account for all the disappearances, while explaining how the survivors managed to continue?

Many explanations of dinosaur extinction have been offered over the years, varying from the possible to the wildly improbable. Some are gradual — with a climatic or geographical or biological cause. But since the discovery of a concentration of iridium in the Cretaceous-Tertiary boundary clay in Italy in 1980, only one catastrophic explanation has dominated media attention. In its simplest form, the suggestion is that an asteroid struck the earth at the end of the Cretaceous. It threw up so much debris that the sun's rays were blocked out for several years until the dust settled. The larger plants lost their leaves, and the herbivorous dinosaurs all died. Carnivorous dinosaurs were able to eat carrion for a while, but then died out too for lack of food. As the dust settled, all the tiny mammals that had been hidden in holes emerged. They took over the world, leading ultimately to us.

Like all simple explanations of complex problems, this scenario is unsatisfactory on many levels. There certainly seems to be evidence that there was a big bang of some kind — not just at the end of the Cretaceous, but also around the time of the earlier dinosaur extinction. But there is much argument about exactly what the event was, and how it may have affected the living world. And there is a strong suspicion, which cannot yet be disproved, that the dinosaurs may have disappeared before the big bang. Exploration of this question has occupied a number of entire books, but we can look at the extinction events as they affected Canadian dinosaurs, and the involvement of Canadian scientists in seeking solutions to these intriguing problems.

A HOLE IN QUEBEC

Although there has been abundant evidence of an extinction event at the end of the Triassic in marine rocks and fossils, at first there was not much evidence that it affected land animals. Part of the problem was that the land faunas were sparse and poorly dated, and it has taken much detailed work to ascertain exactly which different animals lived and when, and therefore what significant changes took place. The earliest Jurassic discoveries in Nova Scotia appear to be only about one million years after what geologists call the T-J boundary; Paul Olsen calls it "the day after."

"The nice thing," Hans-Dieter Sues points out, "is that we have dune deposits, we have lake deposits, we have fluvial river deposits, and all these show the same thing: that all the typical Late Triassic forms are gone, and have given way to a completely different type of community." Many groups of vertebrates known in the Late Triassic appear to have been lost, though the survivors are known in the Triassic. "Dinosaurs appear to have been essentially unaffected by the end-Triassic extinction event. To date, there is not a single recorded disappearance of a family-level taxon of dinosaurs across the Triassic-Jurassic boundary."

By contrast, there is a major change in the pollen record, with assemblages of a diverse plant life being replaced by "greatly impoverished Early Jurassic ones that are almost entirely composed of pollen of certain conifers. In the same rocks further south, the boundary is marked by a "fern spike" — traces of a plant community that spore evidence suggests was ninety percent ferns.

Alerted by the asteroid impact theory for the end of the Cretaceous, Paul Olsen looked for a "smoking gun" for this event. In 1982 he suggested he had found a possible candidate in Quebec, 500 kilometres (310 miles) northwest of the Bay of Fundy. Viewed from the space shuttle, the Manicouagan impact crater is an enormous ring lake, some 70 kilometres (43 miles) in diameter, and despite ice age damage to its features, it has been possible to calculate that it represents an object with a diameter of about 10 kilometres (6 miles) striking with a velocity of 25 kilometres (15 miles) per second, and releasing 100 megatons of energy — ten thousand times the world's current nuclear capacity. Initially the date seemed about right, but more precise techniques have placed it at about 214 m.y. ago — which seems to be too old for the Triassic-Jurassic extinction event at around 208 m.y. Dale Russell comments that: "Although the die-off of reptiles of Triassic Age could have spanned several million years, there is evidence to suggest it was quite sudden." The search is now on for both evidence of extinctions from that date, and another possible impact at the time of the extinctions. Shocked quartz (one sign of an impact) has not yet been found in Fundy, but has turned up in Italy right at the boundary. But so far a crater has proved elusive.

If it wasn't an asteroid, then what other explanations are available for this earlier event? As the Triassic moved to a close, the climate became increasingly hot and dry, which favoured reptiles but slowed down the mammal-like reptiles that were evolving towards a mammalian condition. Perhaps the dinosaurs survived because they were developing a more agile lifestyle, giving them a superior competitive position over their more primitive rivals. But Olsen is comfortable with a catastrophic explanation.

FIREBALLS AND EXPLODING STARS

In the summer of 1972, I was working with colleagues on a dinosaur site across the Red Deer River from where the Tyrrell Museum now stands. Across the blue sky, a white light travelled rapidly, from south to north. It was not a high plane leaving a vapour trail, but an intense spot of white light. We watched it with fascination but had no ready explanation, and since it seemed to have nothing to do with dinosaurs I did not even make a note of it in my field book. It was not until later that I learned that I had observed one of the best-documented asteroid sightings of this century. It had crossed the western United States and Canada, but did not hit Earth; instead, it skipped off into space. Astronomers estimated its size as 8 to 10 metres (26 to 30 feet), and its weight between 100 and 1000 tons, and it came within 52 kilometres (31 miles) of Earth. Less than a decade later, such an occurrence would immediately have been associated with the end of the dinosaurs.

The dinosaurs had survived the Triassic-Jurassic extinction event, but (apart from the birds) they did not survive the one at the end of the Cretaceous. Consequently, every palaeontologist had to have an explanation. Brown's ecological interpretation has already been quoted; in 1935 Parks discussed another, but was rightly suspicious of its simplicity.

> The evolutionary changes in many of the Ornithischia indicate a development of organs of defence against the Theropoda [carnivores]. Did the defence fail? Was the extinction at the end of the period due to the devouring of the Ornithischia by the Theropoda, which in their turn, succumbed to starvation? It is a plausible but dangerous conclusion, for animals of prey do not, as a rule, completely destroy their own food supply.

Dale Russell has taken an intense interest in the extinction question. As early as 1972, he published a brief paper on what he called "The Great Crisis of 65 million years ago." He brought together a wide variety of evidence, and recognized a chemical fingerprint (an unusually high proportion of potassium 40) above the extinction zone in a Montana site.

"Compared to the usual pace of events in earth history, these events seem to have lasted a relatively short time." Russell's search for a possible explanation led him to astronomy, and he charted records of supernovae from Chinese records in 1006 A.D. onwards. These exploding stars "have only recently become the object of close study." One relatively close to Earth could generate intense radiation for a short period, which would have very severe effects on the chromosomes of large terrestrial animals, and produce major climatic disturbances. Russell recognized that the "theory . . . is based on very imperfect evidence," and hoped it would "stimulate interest in the problem."

Russell discussed the problem in other articles, considering a range of possible earth-centred and extraterrestrial causes, and in his 1977 book focuses on the supernova theory. In the same year he convened the first of two K-TEC workshops, named for Cretaceous-Tertiary Environmental Change and jointly sponsored by the Paleobiology Division of the National Museum of Natural Sciences and the Hertzberg Institute of Astrophysics, National Research Council. (K is the geological symbol for Cretaceous, as C has already been used earlier in the geological time scale for Carboniferous.) Of eleven attendees, all but two were from Canadian institutions, an indication of the breadth of expertise available. A prophetic call was made for trace element studies of the boundary layers.

SHOCKED QUARTZ AND SMOKING GUNS

In 1979, two years after the first K-TEC conference, a report by physicist Luis Alvarez and his geologist son Walter showed that iridium levels at the Cretaceous-Tertiary boundary in Italy were more than three hundred times higher than those immediately above or below it. The discovery of this anomaly created enormous excitement among scientists and the public. An explanation was at first sought in the supernova idea. This was predicted to produce high platinum levels, and these were detected in the first sample to be checked. "We were terribly excited," reports Luis Alvarez. Unfortunately his associates were unable to find the same results in a second sample, and they eventually realized that there must have been contamination from previous work with platinum in the same apparatus. And indeed, this finely detailed work is subject to contamination from unlikely sources. In discussion at the second K-TEC conference another apparent iridium anomaly "was on the platinum engagement ring of the person who prepared our samples."

K-TEC II was held in Ottawa in 1981, with twenty-four participants specializing in a number of different fields bearing on the boundary question. Nearly half were from central Canada, while nine (including both Alvarezes) came from the U.S., and four came from overseas (Holland, Scotland, and Switzerland). By this time the iridium anomaly had been reported from seventeen localities, on several continents and in the ocean. None were in Canada, but a section across the Cretaceous-Tertiary boundary had recently been studied in the Red Deer River valley, and Jarzen had been doing careful pollen sampling in Saskatchewan.

The pattern of other rare elements was interpreted as characteristic of a meteor. Three craters, possibly of the right age, had been identified in the U.S.S.R., but little was known about them. Volcanic sources of the iridium, and the possibility that impact craters could be masked by triggering volcanic eruptions, brought special attention to Iceland and India. There was discussion about the marine and terrestrial extinctions, and how it could be proved they occurred at the same time. Different factors that might define the boundary were discussed, and how one could reconcile changes in marine micro-organisms or pollen, the presence of the boundary clay or other stratigraphic boundary, and a high iridium anomaly. Although no conclusion was reached, a note of caution perhaps sums up the discussion. "At this stage," said Dewey McLean of Virginia, "it is wise to consider as many alternatives as possible, and keep examining the data." Dale Russell's final remarks (very significant for dinosaurs, though he was referring specifically to ichthyosaurs) were that "abrupt extinctions occurred in some groups when they were already in a decline."

By 1983, shocked quartz (in which the crystal structure has been deformed by great pressure) had been found at the Cretaceous-Tertiary boundary in Montana and other localities, along with microtektites (small beads of glass that seemed to have cooled in flight after an impact). In his 1989 book, *An Odyssey in Time,* Russell mentions a possible sea floor crater in the Indian Ocean as a candidate, but an underground structure had been reported at Chicxulub in the Yucatan peninsula back in 1981. As glass was found in Haiti and there were signs of a tsunami (earthquake-driven wave) on the Gulf Coast, interest gradually focused on this feature, which was close enough to have blasted glass as far as Montana and southern Canada. In 1989 Canadian geologist Alan Hildebrand (then at the University of Arizona, but later with the Geological Survey of Canada) found a clay layer in Haiti that he felt was impact debris blown out from a crater. This led him to the Chicxulub structure, which he and its original discoverers showed was an underwater impact site. In 1991 borings through the 175-kilometre (110-mile)-wide structure showed shocked quartz, and it was dated at 65 m.y.

"The evidence pointing to Chicxulub has been available for years," Hildebrand said. "The problem was that scientific prejudices favoured other explanations." Some geologists, however, felt that the prejudice was on the side of the catastrophists. University of Saskatchewan's William Sarjeant reviewed a wide range of geological opinion, and showed many scientific reasons for caution. Dinosaurs "declined progressively during the Late Cretaceous," far too many groups other than dinosaurs had survived for a catastrophic explanation to be convincing, and there was no agreement on the actual boundary. A "Great Dying" at the end of the Cretaceous, Sarjeant suggested, was "more a consequence of determined faith than investigative science." In a more recent review he points out that the available sites "fall within a quite tightly circumscribed area of North America." However, to some scientists and many of the public there seemed no reason to doubt that the "smoking gun" responsible for the end of the dinosaurs had been found.

AN OPEN VERDICT

At the KTEC-II conference, Russell explored the question of whether the dinosaurs were reduced in numbers before the extinction point. In Montana he and others had noted "a separation of several metres between the highest articulated dinosaur bones, and the horizon . . . usually associated with the Cretaceous-Tertiary boundary." This thickness of rock might represent tens of thousands of years. As the evidence for the impact grew, increasing numbers of vertebrate palaeontologists pointed out that the dinosaurs may have become extinct before the big bang.

The best evidence of dinosaurs near the boundary is in Montana and Alberta. In places in Alberta, Cretaceous beds seem to grade without a break into the overlying Paleocene, so the boundary can be examined in detail. A critical question is the accuracy of the dinosaur data. Arguments are made about sampling errors (Are dinosaurs absent or have they just not been found — or even looked for?), statistical validity (Is the decline real or apparent?), and interpretation (Are dinosaurs recorded to family, genus, or species?). The most significant problem is the assessment of negative evidence. While a fossil dinosaur bone is a positive indication of a once-living dinosaur, does the absence of a dinosaur bone mean that there were no dinosaurs living at the time? Clearly this cannot be so in every case, for many gaps in the fossil record obviously have been bridged by animals that must have been living somewhere.

In Alberta, dinosaurs seem to have been declining during the last two million years of the Cretaceous. A comparison of the Scollard

Formation, closest to the boundary, and Dinosaur Park shows only one-quarter of the carnivores, and half as many plant eaters. There are now about thirty Cretaceous-Tertiary boundary sites in Canada (about as many in the western U.S.).

In 1993, spring runoff created a slump of a 50-metre (165-foot)-wide block of prairie, producing the best exposure of the Cretaceous-Tertiary boundary in Canada. Dennis Braman, palaeobotanist at the Tyrrell Museum, reports on the plant evidence:

> From half a metre to 5 m below the boundary (depending on the section) we have a good assemblage which is widespread and rich in [plant] species. Then there's an impoverished zone, with some of the same species, well-preserved, but poor in numbers. It may be a coincidence, but the last dinosaurs are two, three or four metres below this zone. Often another angiosperm will replace the species below — their grains have simpler ornamentation, so they may not be insect pollinated.
>
> The boundary claystone is subdividable into three layers. The lowest hackly layer has no iridium, no shocked quartz. Next is a darker-coloured satiny layer, where iridium and shocked quartz start. The gymnosperms get cut down, and there's a fern spike. Fern spores have been relatively uncommon before, but they come in good numbers — up to 80% of the total pollen and spores. The upper layer . . . has the fern spike and shocked quartz as well. The coal above — which is found over a very wide region from the Northwest Territories to New Mexico — may also have ferns, and the gymnosperms gradually increase again.

His conclusion? "The geochemistry supports the impact, but the environments were changing anyway. It's hard to say that the impact killed the dinosaurs."

Canadian palaeontologists, like those elsewhere, are divided in their views. At one extreme, Sarjeant considers that "any extraterrestrial impact is quite as unlikely a cause of dinosaur extinction as is nuclear attack for the extinction of the mammoths." Currie sees "a gradual reduction in numbers of species of dinosaurs, although evidence for something major happening at the end of the Cretaceous is also pretty robust." Dale Russell is concerned with methods: "You cannot study why the dinosaurs died by studying dinosaurs. . . . There are three hundred eleven fragments of dinosaurs, worldwide, for the last nine million years of Cretaceous time, and you have billions of atomic nuclei which you can examine, like pollen grains, soot and shells." He recognizes an apparent reduction in dinosaur numbers, but feels that "specimens are evidently as abundantly preserved" in the "relatively limited exposures of the Scollard Formation." Both Russell and gradualist Peter Dodson agree that "the latest Cretaceous dinosaurian record is far too incomplete to support either the catastrophic or the gradualistic model in a statistically meaningful manner."

Is an answer to the extinction question possible? At present it is hard to say. But the exciting convergence of data from astronomy and geochemistry, from articulated skeletons and tiny pollen grains, offers the promise that some critical mass may soon be reached, as with other major breakthroughs in earth science in recent decades. Until then all we know for certain is that while many life forms (including those dinosaur-descendants, the birds) continued into the new world of the Tertiary, the rest of the diverse animals we know as dinosaurs did not make it past — or even to — that boundary. As Dale Russell reminds us, "Majesty, proportion and beauty were diminished."

DALE AND THE DINOSAUROID

Suppose the end of Cretaceous extinction hadn't happened? What would the dinosaurs have become? Dougal Dixon's entertaining speculation, *The New Dinosaurs,* gives his views of the way dinosaurs might have evolved in North America. The elaborately horned "sprintosaurs" evolve from the crested duckbills; the "monocorn" develops from an ancestry in the horned dinosaurs. The "balaclav" is a specialized hypsilophodont, which has "evolved insulating layers of fat and fur" and lives in alpine regions.

Dixon's small carnivore descendants include a group of "arbrosaurs," adapted for a life in the trees; the "mountain leaper" and the "northclaw"; and a little carnivore that "apart from its large head and furry coat," is mainly notable for "the massive single claw, the killing organ on its right forelimb." The "springe" lies in wait for pterosaurs, smelling of putrefaction, and impales its victims with its killing claw.

The little carnivores have given rise to more serious speculation in the creative mind of Dale Russell. His early interest in the small carnivores (see Chapter 8) led to the discovery of a partial skeleton of *Troodon.* "Well, isn't this interesting," was Russell's reaction. "It had big eyes. They were somewhat focused ahead. I had its wrist, and I knew that it had a tridactyl hand. Obviously it had a large brain, and it's bipedal, and what does this remind you of?" Dale recognized the similarity between the little dinosaurs and the ancestral humans — and in due course enjoyed showing the bones to palaeoanthropologist Richard Leakey.

Russell had been consulting with NASA's Search for Extraterrestrial Intelligence program, and had made studies of the evolution of the human brain. He was working with artist Ron Séguin to produce a model of *Stenonychosaurus inequalis* — now *Troodon formosus* — (which appears in the photograph of the attendees at the K-TEC II conference). They continued to explore what Russell called "a thought experiment" and jointly published the results in 1982. "It might . . . be entertaining," he explained, "to speculate . . . on how the descendants of *S. inequalis* might have appeared had they survived the terminal Mesozoic extinctions, and achieved an encephalization quotient similar to that of *Homo sapiens.*"

With careful reasoning Russell traces the potential development of the skull to house a larger brain, the extension of bipedalism, and the development of a grasping hand. Compared to its dinosaurian "ancestor," Russell's "dinosauroid" is very human in appearance, standing upright. The three-fingered hand and reptilian face are the most obvious external distinctions.

The dinosauroid has received some attention in scientific books on dinosaurs, but — once created conceptually — has gone on to have a life of its own in speculative fiction trilogies written for both children and adults (see Chapter 12).

Robert Bakker went further, and considered how our ecosystem would be if Russell's scenario had happened. "You and I, dear readers, would probably be members of some tiny species, eking out a terrified living under the ever-present shadow of a dinosaurian overlord."

10

CRETACEOUS PARK

"I never knew how long I slept, but when I awoke, I was overcome with surprise, I could not tell whether I had awakened in eternity, or Time had turned back his dial, and carried me back to the old Cretaceous Ocean."

CHARLES H. STERNBERG

RECREATING THE PAST

Labouring during the day to discover, expose, and collect the bones of dinosaurs, Charles H. Sternberg perhaps dreamed at night that he saw his discoveries while they were still alive. Certainly in his 1917 book, *Hunting Dinosaurs in the Bad Lands,* he strove in some of the later chapters to carry the reader in imagination back to the past. He describes the landscape and vegetation as he envisaged he himself would have seen it if he had travelled back in time, from the perspective of his own experience as a dinosaur collector. Living a sort of Robinson Crusoe existence in the past, he climbs trees for safety, collects shellfish for food, and makes a dugout canoe.

Along the way he encounters in imagination the creatures he had collected in his later, real existence. He finds turtles and garfish in the streams, and: "Soon, I saw the white foam ripple off the huge back and tail of a swimming reptile, a duckbill if you please, that was rapidly approaching." He describes its behaviour, and watches it being attacked by a carnivore. In later chapters of his fantasy he is joined by other members of his family, and visits the marine Cretaceous of Kansas, and then the earlier Permian period whose rocks he had studied in Texas.

Sternberg was a collector more than a scientist, and did not have a reputation to lose by being too fanciful. Later scientific writers have been much more cautious. Lambe and the younger Sternbergs are generally prosaically scientific in their language, and although Parks suggested that "We can picture many a chase, many a combat, and many a death-cry disturbing the silence of the central sea," he did not describe the events he pictured.

More recently it has been realized that storytelling is of potential use in identifying some of the questions scientists must ask, as well as a legitimate tool of interpretation for the non-specialist. "Speculation (= scenario building, story telling, creative logic, etc.) is the easiest, most colorful, most untrustworthy, most maligned, but sometimes most fruitful approach," says Walter P. Coombs, Jr., discussing techniques for understanding fossils in a recent text.

Palaeontologists are now more willing to use the approach of imaginative recreation, and are less likely to be criticized for it. Phil Currie asks us to "Imagine sitting at the edge of a forest and looking across a marshy plain dotted by horned dinosaurs feeding on ferns and cycads, preparing for their annual migration north into the Arctic Circle. There is no grass, but most of the other plants look astonishingly modern. As you listen to the singing of the birds in the trees and bushes, a shadow races across the ground, and you glance up to see a pterosaur the size of a small airplane soaring overhead!"

In his books, Dale Russell gives the main burden of reconstruction to the fine paintings of Ely Kish, but his words also help us visualize past situations. In his 1977 *A Vanished World,* Russell describes the "cow-like majesty" of *Edmontosaurus,* and suggests possible behaviours of other dinosaurs in brief, pithy phrases. His later book, *An Odyssey in Time,* also contains vivid word pictures, as he pictures the origin of the Dinosaur Park *Centrosaurus* bone bed. "A *Centrosaurus* herd attempted to swim across a large stream, but because of panic, current speed, or some other reason, a large number of animals drowned. Hundreds of carcasses were stranded along the banks of the river, where they putrefied. Many hundreds of tonnes of decaying flesh attracted scavenging tyrannosaurs. They trampled the skeletons, breaking bones, and cutting grooves into them with their teeth."

Sternberg's duckbill is swimming, for this is the way they were interpreted when he wrote early in the century. And as we saw in the previous chapter, Russell's interpretation of the *Centrosaurus* bone bed might also be replaced by different views. Thus, everyone's reconstructions are dated; the more closely they reflect current scientific thought when they are written, the more they seem "old-fashioned" when new evidence modifies that interpretation. As more evidence is gathered and theoretical questions are explored, there is always room for new attempts to visualize the ancient world.

TRAVELLERS IN TIME

Let us, through the power of imagination, take a trip back in time, to the zenith of the Cretaceous, as reflected by the wonderful diversity of landscape and life in Dinosaur Park and surrounding areas. Through the imaginative literature of time travel, vehicles are available to take us there, whether we choose the intangibility of a dream, or the fantastical mechanics of H.G. Wells' *Time Machine.*

How should we prepare for our trip? Charles Sternberg dreamed his way into the Cretaceous with only his field bag. Dale Russell offers more practical suggestions for the Cretaceous traveller: "If you must go, dress in a long-sleeved shirt and trousers, and wear a mosquito net round your hat, for biting insects will be abundant. Put a knife, machete, cord, hammock and mosquito netting in your knapsack, as well as mosquito repellent and matches." So far, the equipment is not too different from that of a modern dinosaur hunter, but Russell goes on to recommend "a high-powered rifle and several hundred rounds of ammunition." Suddenly, the large carnivores become a little more real, especially when he warns us that the rifle is only useful for small carnivores, for "It would certainly be futile to attempt to shoot a tyrannosaur . . . their brains are very small and are well protected by bone." Russell advises on safety precautions ("hang your hammock at least 25 feet above the ground"), suitable food, and choice of habitat. He even envisages the possibility of training a *Stenonychosaurus,* as the only creature "that could, to any degree, fill the role of a dog."

If our visit is not to be preoccupied with survival, we need to be more self-effacing. Let us suppose (borrowing another of Wells' ideas) that it is possible to be present in the landscape, but invisible. Then we can wander around without attracting the attention of a hungry carnivore — providing we can insulate ourselves against their sense of smell.

Birders — those observers of modern dinosaurs — have their Christmas Bird Counts, or go out for a spring "big day," trying to see as many species as they can within a twenty-four-hour period. To really enjoy Cretaceous Park, we need a big day in the past, a chance for a small group of friends to have fun, sample a number of habitats, see what is there, and try to understand a little of the ecology of the time and place. It's what I might do in a new area in preparation for leading a field trip for a museum, university, or natural history society: noting the most interesting spots and best times to be there, and making field notes so that I can check later what I saw, and looking up things I am at first unable to identify.

My introductory talk to groups would go something like this: We are taking a single trip — expect a big day — but we won't see everything that we know lived in the area. And since our knowledge of past animals is largely from bones, we won't be able to identify everything firmly — we don't yet have a field guide. We have chosen late spring, so there is a chance of seeing some breeding activity. Our timer is set for about 75 m.y. ago — in the middle of Cretaceous Park's most diverse life. If we went at a different time — even during a different season — we would see different things. And on a real trip we would surely see things that we do not recognize; although the fossil record is so rich, ongoing new discoveries show that it is incomplete. We'll look at dry upland areas, the wetter areas where Dinosaur Park now is, and then travel east to the seashore.

Bring binoculars and a camera — and plenty of film! While we are there we will travel on an airsled that can hover and glide without touching the ground, which will help get us from place to place, and will carry our supplies. It is covered, so we have some protection from the weather, and our scent won't be carried to predators. Its built-in compass is adjustable for different magnetic regimes, so we can keep track of our direction. Its built-in radar will help us avoid collisions — the only way we can affect living things.

We know we must not eat, drink, or touch anything we see, and certainly not injure or kill anything. Even more than in a visit to a modern park, we must take only photographs, and if possible, not even leave footprints — after all, they might be fossilized. The consequences of real interference might be serious. If we foolishly allow ourselves to be eaten by a dinosaur, this will not only increase the time-travel insurance, but we might give the animal a competitive advantage that changes the course of dinosaur evolution. Were we to step on a small mammal, we might even change our own future, and return to the present to discover that we humans had never evolved.

So — cautiously — come with me to the Cretaceous. . . .

SQUAWKS AT DAWN

It is dark, cool, and silent. There is a faint hint of lightness on one side, a vague sense of odd-shaped trees looming close by. As we sit listening, we become conscious of intermittent faint rustling in the indistinct vegetation. Gradually, the eastern sky glows yellow, then pink. High clouds catch the glow, and the trees become more distinct. Each has a crown of long feathery leaves, like giant ferns — these must be cycads.

The woods are silent; we cannot expect much of a dawn chorus here. Suddenly a loud ringing call echoes in the near distance, but (even with night glasses) we cannot see what has made it. Another comes from further away, then more, until perhaps a dozen creatures are squawking. We cannot see what these are, but suspect birds or other small dinosaurs, which with their insulation of feathers are more likely to be active at this time of day.

The rustling has continued; obviously, small mammals are foraging in what we can now see to be a dense patchwork of ferns, clubmosses, and horsetails that cover the ground. Occasionally the tip of a horsetail waves gently as if in a faint breeze, then stops. It is still gloomy beneath the trees, when we become conscious of a more purposeful movement in the ferns. Although we get only occasional glimpses of a small head on a long neck, about a metre (3 feet) above the ground, we can follow the movement of what is obviously a small dinosaur by the disturbance in the plants. It suddenly slows and emerges into a more open area, purposefully stalking something we cannot see with slow steps and alert angling of the head. The neck suddenly extends, its long tail flourishes in the air, and there is a single squawk. We get a brief glimpse of its little head struggling with something small and furry. In the ensuing silence, we follow its more deliberate progress into the forest beyond, and realize that a little *Troodon* will not go hungry today. As it disappears, we realize that, although it is neutrally coloured, its surface seems to have a fuzzy appearance — a covering of proto-feathers.

As the sun suddenly pierces the gloom, we turn to survey the landscape. We are on the break of a gentle slope, in a gap between groves of trees. The clump of cycads downslope is between us and the growing light, and other clumps of trees on low hills block off our view to the north and west. We can see that some of their branches are broken and show signs of browsing. Although the undergrowth is dominated by ferns, we can see some flowers among them — lilies, perhaps — and a magnolia tree is in bloom. Spider webs sparkle with dew.

Between us and the sun, a wide, low plain extends to a far distant shore, where faint sparkles suggest the sea begins. To the northeast, lines of trees suggest a series of rivers flowing from the slope. Clumps of other trees and occasional gleams of water show lakes with wooded shores on the plain — the area that is now Dinosaur Park. In the far distance we can see that the rivers seem to reach the sea in an estuary.

As the light grows, we see that plants do not flourish everywhere. A wide swath of trampled vegetation comes from the plains and passes us to head into the trees. We edge the sled over for a closer look. Some of the wetter patches show footprints that — although we have seen similar fossils — appear astonishingly large when freshly impressed in soft mud. The tracks show the three broad toes characteristic of duckbills; most are large, but some are of smaller sizes. Many tracks are worn and seem to have been made some time ago, while others are very recent. In the low, angled light, they show up well, and our cameras are soon busy.

Along the lines of tracks, bulky lumps of what is obviously dinosaur dung are scattered. Older pieces are decaying, showing scraps of what look like pieces of stem of some kind of plant. On a fresh lump we notice large beetles, round and black with a purplish lustre. One beetle gathers dung, and another rolls a ball towards a burrow in the soft ground beside the trackway. We have already seen a dinosaur as predator; here is a hint of the natural community that depends on the dinosaurs for food and habitat.

THROUGH THE FOREST

We start up the sled and glide slowly along the broad track that climbs over bumpy ground into the forest. The trees include cycads, but they become more varied as we climb. There are some firs of unfamiliar types, and some broad-leaved trees, whose leaves are just springing. The trees are snarled and tangled — some appear to have blown down, and they are connected by trailing lianas into a tangled mass. On the more level branches and contorted trunks, flowering plants — epiphytes — are beginning to bloom. Soon we move into a stand of very tall, stately redwoods — even though the sun is already striking hot on our backs, it looks cool and shady underneath the trees.

Suddenly, a lanky ostrich-mimic strides across the track, and we stop to watch it ambling away through the fallen branches and sparse ground vegetation under the redwoods. Before it has disappeared, a buzz of flies alerts us to something sticking out into the trail. It is a tail, which leads us into a clump of young ginkgoes to the carcass of a large but slender dinosaur, clearly one of the duckbills. Its dull greenish body is badly torn, with one leg missing and the neck badly mauled. A couple of ostrich-mimics are slobbering in its entrails, while one reaches its long neck into the body cavity in search of some other delicacy. Its neck looks to be broken and torn, and a long, bright-red crest angles back from its head — it looks like *Parasaurolophus.*

One of the larger carnivores obviously made a successful kill, perhaps last night. As soon as we realize this, we start to look around nervously, although we know we are invisible and the enclosed sled confines our scent. We are close to one of the conifers, and one of our party points out the rich resinous sap oozing from a break in the bark. Ants are dashing busily up and down the trunk, and one of them may well get caught in the resin and end up in a future piece of amber.

Continuing along the trail, we see the forest gradually opening up ahead of us. Occasional glimpses show us that there are birds and pterosaurs over the forest, but we have no clear views. As the trees fall back on either side, we see fresh sign on the trail — steaming piles of dung — and hear distant hooting ahead of us. We slow down, and cresting a rise, we see ahead of us a small herd of hadrosaurs on the march.

Parasaurolophus *shows its long crest.*

Slowing to their walking speed, we see a mixed group: three *Parasaurolophus*, and nearby a larger group of crestless duckbills with prominent Roman noses — *Kritosaurus,* perhaps? — coloured in a striking variegated pattern of dark greens. They are all ambling along steadily, using all four legs. Occasionally one pauses to rear up against an isolated tree and pull down some fresh leaves. We have already speculated that the heavily used trail may be a migration route; here we seem to have a group on the march, heading north towards the Arctic to take advantage of the abundance of vegetation produced during the long summer hours. Suddenly, the paras stop; the largest and smallest turn towards each other and begin nodding their heads, their scarlet crests flashing in the sun. Now we see where the resonant hooting call comes from; this is a pair-bonding ritual of some kind, helping the animals to stay together on the long journey. The third animal also hoots a little, but half-heartedly. Perhaps it was his mate whose body we saw back in the forest — if so, we cannot but hope he will find another one as he continues the long journey.

TO THE DRY PLAINS

Cresting a rise, we pause to survey the scene. There is higher, drier ground to the west, with mountains in the distance. Some of them are obviously active volcanoes — plumes of smoke rise and blend with high clouds that have been gathering. Some of the ground and vegetation is grey; even this far from the mountains there has clearly been an ash fall since the last rain.

What looks like a large, dark bird circles over the plains; as our eyes get used to the distance, we realize there are others further away. Binoculars show us the smooth wing outlines of a pterosaur. These are big, but we can't guess the size without anything to compare them with. As we watch, the nearest one swoops into a long glide, and two others move towards it.

We head to a rendezvous with the gliding pterosaur. The ground is sparsely covered with mosses and clubmosses, and bare patches begin to appear. We are moving into a drier area, perhaps in the rain shadow of the mountains. Tall anthills rise from the dry ground. We pass occasional lakes, the larger surrounded by small patches of trees, the smaller crusted with minerals. By one of the larger lakes we see our pterosaur, settling beside a lump on the ground and folding its huge wings. A scurry of small carnivores backs away as we get closer and realize that the pterosaur is nearly as big as the small dinosaur lying before it, dead and decaying on the ground. Here we must have *Quetzalcoatlus,* the biggest pterosaur known. Using the hand on one of its wings to pull aside a patch of loose skin, it reaches its long bill and neck into the body cavity, and pulls out a trail of intestines. As soon as it is preoccupied, the small carnivores begin to approach again, and cautiously tear at bits of the carcass out of reach of the flying monster. They are more solidly built than the *Troodon* that we saw earlier, and as they tear and squabble we realize they have big claws on their centre toes, held high off the ground — dromaeosaurs. They may be scavenging, too, or the pack may have combined its efforts to destroy this victim, swarming it like a pack of reptilian wolves.

What is the carcass? We circle around until we can see its head. Although there is damage, we can see a domed crown, circled by blunt spikes. Here is a small pachycephalosaur, a bonehead that might be *Stegoceras.* We imagine the kill with the group of dromaeosaurs attacking it from all sides, while it could only defend itself by charging one enemy at a time with its battering ram.

As two more giant *Quetzalcoatlus* soar in, looking like small airplanes, the dromaeosaurs scatter again. One of them is limping badly; our scenario looks very credible as we realize it might have got too close to the pachy's hard head. We leave the animals to redefine the pecking order around the carcass, and turn north.

DOWN DINOSAUR RIVER

Ahead of us, the ground falls gradually, and trees become more frequent. Dinosaur trails thread between the trees, and we see more duckbills heading northwest. Soon, we are looking into a broad, shallow valley, filled with an open forest. Through the trees we catch glimpses of several rivers, flowing from the western mountains, and off to our right. We pick our way down the slope, through an open patch where young trees are greening up the site of a former forest fire, until we reach the first river. We turn east, skimming along a few feet above the brown, sluggish water. The river curves gently, muddy banks with abundant tracks on one side and eroding slopes on the other. Its banks are lined with tall, broad-based swamp cypresses, some rooted in the water and some collapsing into the stream where the bank is eroded beneath their roots. Behind the cypresses, the trees are a rich mixture of conifers and broad-leaved species, hung with creepers and trailing sheets of lichen. We are in the area that in our time is Dinosaur Park. Huge dragonflies hawk for mosquitoes along the banks.

We pass a battered tree trunk drifting in the water, and see others stranded on a sand bar, where the river divides to pass a richly wooded island. We slow as we realize that the bumps on the stranded logs are huge turtles, each a metre long, enjoying the bright sunshine. Then we realize that a substantial crocodile is lying on the sandbank too, pretending to be just another log.

There are open patches among the trees — old burns, shrubs broken by feeding animals, and a swath of crushed young trees where we glimpse a tank-like armoured dinosaur pushing its way between two clearings. These spaces give us glimpses of vivid fungi and pale flowers, hawk-like birds (*Apatornis*, perhaps), and small pterosaurs. An opossum moves slowly along a branch, using its tail to help it hang on.

A large, crested duckbill is perhaps *Corythosaurus.* We are tempted to watch it, but there is an overpowering smell of animal decay. Buried in the rocky shore beside us, we distinguish the outline of the carcass of a large armoured dinosaur, upside down with one broken leg bone sticking up pathetically from the stones. Its armour has clearly discouraged large scavengers, but our noses tell us that the bacteria are doing their job, and we move on with relief.

Around the next corner, we stop to watch a half dozen hadrosaurs, the biggest around 15 metres (50 feet) long, tear into a tree. They are crested, and we decide they are probably *Lambeosaurus.* Standing stolidly on hind legs, they reach up and pull branches into their mouths, and rip off the leaves, dropping debris around them. The group includes animals of different ages, and their crests are not all the same. They hoot gently to each other as they are feeding. Eventually the biggest one approaches the river, looking cautiously before venturing to swim slowly across. The others follow, a young one last. As we watch the leaders clamber out on the far bank, the young one stumbles, then surges forward, splashing violently. Amid agitated hooting from the others, we see it pulled back into the river by a 4-metre (13-foot) crocodile, which obviously has a firm grip on one hind leg. The struggle is brief, and by the time the young lambeosaur has been drowned, the large leader is already heading off into the forest, followed by the others.

We continue downstream, hearing squawks of birds and perhaps small dinosaurs from the forest. We see other duckbills, but cannot identify them. The river widens, winds more, divides and combines round many islands, and we see many basking crocodiles. One sandbank is littered with bones — a bone bed in the making. The areas behind the levees become lower and more swampy, with abundant cattails and giant horsetails. A lone horned dinosaur — could it be

Chasmosaurus? — is up to its knees in a slough, uprooting cattails and munching on the roots. Slimmer crocodile-like reptiles that we take to be champsosaurs float in the lagoons.

UP FROM THE SHORE

Another river joins from the north, and the wider waters slow. There is a porpoise-like flurry in the water, and a head briefly emerges — a fish clutched in its mouth — on the end of a long neck. Fossil elasmosaurs — swimming reptiles — have been recorded in river deposits, so we are not surprised to see it. As we continue east, a line of debris along the bank shows that we are entering tidal waters. We cut off across a long spit and head south along the coast, where wide sandy beaches alternate with muddy lagoons, ringed with cattails and low trees. There is a sudden crack of thunder and we realize the clouds in the west have been building while we headed downstream; soon we find ourselves buffeted in an intense storm. With this low shore, a bigger storm with an onshore wind could back up the rivers and drown this flat plain.

In the curtains of rain a heavily built armoured dinosaur munches stolidly on low-growing vegetation. Black and white gull-sized birds — *Ichthyornis*? — sit on the beach or dabble near the shore. Out to sea, dark, loon-like birds, probably *Hesperornis*, are fishing in the waves; the rain makes no difference to them.

The rain stops as we slowly cruise along the strand line of seaweed and broken branches. Clams, oysters, and brightly coloured and patterned ammonite shells — some whole and bigger than dinner plates — litter the beach. Decaying fish and occasional bones lie in the weeds. Further along, a couple of tall ostrich-mimics are picking their way along the line of debris, bending occasionally as they pick up something edible. Bones of a 2-metre (6-foot) sturgeon have been well picked; a bulky carcass looking like a slim, stranded whale as we approach proves to be a small mosasaur, its white teeth gleaming in the afternoon sun. A group of small dromaeosaurs is busy pulling at the carcass.

Behind the shore, the land rises only gently, with lakes and woods alternating with open ground. Spring floods have obviously spread fresh mud over the lower areas, but this is dry now, and young vegetation is bursting out. Back towards the low hills where we watched the dawn, we can see herds of something large, and we turn inland to investigate. We follow a tree-lined abandoned river channel, which makes a series of linear lakes, studded with yellow water lilies. Patches of vivid green mosses mark the boggy areas between the lakes. A yellow scum must be abundant pollen from the nearby trees. A metre (3-foot)-long terrestrial turtle munches stolidly on ground vegetation.

We swoop down to investigate a pair of elephant-sized duckbills dabbling in the water. They are flat-headed, but bulbous-nosed — perhaps *Gryposaurus* — and their plain grey-green colour is startlingly decorated by bulbous skin bladders on their faces, with bright yellow and blue patches. These inflate and then collapse, with the discharging air making the loud honks.

A rainbow forms as the sun breaks through again, now well to the west of us. The herds we glimpsed before the storm were horned dinosaurs, and we approach and circle them for some time. Many seem to be *Centrosaurus.* Some are marching purposefully to the northwest along a network of trails, while others browse, or drink at the sloughs. We move on to observe a smaller group of multispiked *Styracosaurus* munch on shrubs bordering a wood beside one of the lakes, the largest ones occasionally flashing their strikingly patterned head shields at the others when they get too close. Two large ones briefly lock horns and wrestle, each trying to turn the other's head, until the smaller one backs away.

Suddenly, a movement in the wood attracts their (and our) attention. With a silent determination, a large carnivore bursts from the shadows, and heads into the flock. Heads tossing, they panic and run, at first scattering, then herding up again. The carnivore's dappled pattern was obviously good camouflage while it lurked in the forest; now it is all too visible as it pounds after an older, limping styracosaur. From alongside, the huge jaws seize its neck, and as the horned dinosaur falls, the carnivore rolls with it, maintaining its grip. Within moments the styracosaur is still, its underbelly ripped open by the great jaws. A couple of larger styracs slow and stand, threatening, but when another large carnivore emerges from the wood they back away. The newcomer is followed by a couple of young carnivores, and within moments they are all tearing at the carcass. They are quite heavily built — perhaps this is *Daspletosaurus.* Like many predators, they seem to have produced their young earlier than their prey do, so that they are growing when the maximum food supply is available. The attack from sheltered woodland is energy-efficient, and like modern predators they go for the most vulnerable animals — sometimes a young one, sometimes one that is injured or ageing. Perhaps with more warning the styracs might have mounted an effective defence.

We leave the carnivores to their gruesome meal, and head towards the low hills to the west.

A HUDDLE OF HYPACROSAURS

Between shallow forested valleys, low ridges begin to rise — outliers of the hills where we began our journey. On one of these we spot some activity, and turn to investigate. Dozens of medium-sized duckbills are gathered, some moving, some bickering noisily, some sitting quietly. The sitting animals are regularly spaced — suddenly we realize we have found a nesting colony. The ridges on skulls and spine show us that these animals are *Hypacrosaurus.* There are dozens of nests, some protected from the sun by an adult, each sitting on or beside a scrape in the bare earth. Untidy piles of dirt and vegetation show where other nests are being kept warm while they have been left unprotected for a while. Other hypacs wander through the nesting colony, while stationary duckbills lunge and squawk at them. Around the colony, adults are moving to and fro, seeking food or nesting material. A few carcasses show that the colony attracts predators. The most noise comes from one side, and we ease the airsled over to investigate. Several slim bird-mimics are being harassed by the residents, while nearby a metre (3-foot)-long lizard has dug into an outlying nest and is wolfing down an egg. We see the rest of the melon-sized eggs, lying in a double row in the loose dirt. We cruise the nest

Model of a Hypacrosaurus *nest.*

colony, listening to the babel of squawks. One pair is preoccupied with their nest, and we see several 10-centimetre (4-inch)-tall young running about, while other eggs are hatching.

As we glide gently uphill to our starting point, the sun is setting behind the wooded hills. In the small lakes, we can hear frogs starting their evening chorus, and clouds of mosquitoes hover in shady corners. We glimpse a cat-sized mammal briskly crossing open ground, and know the little *Troodons* we saw at dawn will soon be about again.

It is time to leave for what we consider the present day. There are no tourist shops here, we have nothing to take back but our memories and some vivid images; some on film, and some in our heads. But the next time we look at a fossil bone in a museum, or in the badlands, it will conjure up for us a more vivid picture, as we remember our trip to Cretaceous Park.

11

EXPEDITIONS AND EXHIBITIONS

"It began with a simple question. 'If you could search for dinosaur fossils anywhere on earth, where would you go?' Dinosaur scientist Dr. Philip Currie answered without hesitation. 'The Gobi Desert.'"

JOHN ACORN

"YOU NEVER KNOW WHERE HE'LL POP UP NEXT"

Arriving late at night at Vancouver airport in July 1998, I was delighted to spot a photo of a smiling Phil Currie on the front cover of the Canadian edition of *Time* magazine. The accompanying story, "Dinosaurs of a Feather," was about his recent work on feathered dinosaurs from China. Currie was also featured in the *National Geographic* for the same month, and in December was named as one of *Maclean's* Honor Roll of Canadians of the Year. One of Currie's frequent collaborators, Kevin Padian of the Museum of Paleontology at Berkeley, California, says, "Phil is laid back and easygoing, but that attitude masks a tremendous industry. You never know where he'll pop up next."

I had been following the evolving story of the relationships of birds and dinosaurs and had already heard of the new finds, so I was pleased — but not particularly surprised — to see Currie on the cover of *Time*. But it did make me reflect that back in the 1970s when we worked together at the Provincial Museum we would both have been pretty startled if he had popped up on the cover of *Time* — let alone reporting on a specimen from China. At that time we saw exotic expeditions as an activity possible only for wealthy museums in other countries — the U.S. and those in Europe. How had Canadian dinosaur palaeontologists moved — by rather imperceptible stages over a quarter of a century — from cautious dinosaur research on Canadian dinosaur faunas to international recognition for expeditions around the world?

THAT AFRICAN TRAVERSE

True, earlier Canadian dinosaur palaeontologists worked in other countries. Born in Canada, William Matthew found the United States so satisfactory that he never returned home. But the first Canadian to lead a dinosaur expedition far afield was British-born, Canadian-raised William Cutler.

After the First World War, Cutler resumed dinosaur collecting in Canada. He sold fossils to the British Museum, and tried to talk them into a more exciting project. A 1922 letter to Arthur Smith Woodward ends: "I do wish you could send me on that African traverse and I hope that you will yet perhaps be enabled to do so."

The expedition Cutler was angling for was to the classic Jurassic site of Tendaguru. From 1907 to 1912 the German East Africa site was extensively studied by German scientists, who shipped 250 tonnes of bones of huge sauropods and other dinosaurs to Berlin. Following the war, Tanganyika Territory became British, and the British Museum (Natural History) had a chance to try its luck. Woodward found a grant to finance an expedition, but the museum had no one with expertise to lead it.

In 1924 the British Museum hired "Professor" Cutler (who "had made a great name for himself as a collector of fossil reptiles in America . . . and was acknowledged to be one of the very best field collectors of the day.") "Brought up in a hard school, Cutler prided himself in going anywhere with a minimum of kit and the simplest of food, and it came as something of a shock to him when he was equipped for the East African Expedition to find out that his impedimenta contained tents, camp beds, mosquito nets and the like."

A young assistant was hired to provide local knowledge and translation services. The African-born Louis Leakey, then a Cambridge student on his first expedition, was in awe of Cutler at first. He remembered that Cutler "was an excellent teacher." However, Cutler was unnecessarily secretive, and tried to keep the mysteries of plastering bones to himself.

Cutler spent two months collecting shells before following his assistant to Tendaguru, and once there, his attention was everywhere; his consuming interest in butterfly hunting led him to collect over seven hundred specimens in three months. Cutler was also sadistically cruel to the animals he collected, and was irritated by Leakey's confidence with staff and nature, so leader and assistant did not get on well.

Cutler was unwilling to use Leakey's local knowledge and failed to take the precautions Leakey recommended against disease. After Leakey had to return to Cambridge, Cutler remained at the site until he fell ill of blackwater (or malarial) fever and died in August 1925.

Cutler's death at forty-two robbed him of much of the credit that might have come from his work. However, he made some notable dinosaur finds, and his training helped Leakey to become internationally famous as a palaeontologist and anthropologist.

FROM TE HOE TO TENDAGURU

More recent immigrant Canadian scientists have continued work in their countries of origin. William Sarjeant of the University of Saskatchewan had already worked on dinosaur tracks in the U.K. before he came to Canada in 1972. Here his work expanded into many other areas, but he has continued to make important contributions, particularly into the history of British dinosaur trackway studies. Hans-Dieter Sues of the Royal Ontario Museum continues research in Germany, his country of origin. "People who study dinosaurs often have the opportunity to travel," says Dale Russell, while discussing the creation of artistic restorations of dinosaurs. However, in Canada only staff of the Royal Ontario Museum — founded substantially with Middle Eastern antiquities — had a tradition of international research. Such travel generally had not been possible for Canadian palaeontologists, but shortly after Russell joined the National Museum in 1965 Canadian museums began a period of relative prosperity. In 1967 — Canada's centennial year — many Canadians gained a renewed interest in their history. In the west this led to the creation of new provincial museums; for the national institutions new and more ambitious research programs became possible.

Russell was ready to take advantage of such opportunities. Although work on Canadian dinosaurs continued as a primary concern, his interests were international. The small raptors that fascinated him were found elsewhere in North America, as well as on other continents, and data on broader questions of dinosaur extinction and the evolution of intelligence came from many other countries. As he began to commission paintings and models to illustrate his books on Canadian and North American dinosaurs, Russell realized that the best models for their environments are now found in many other parts of the world. "Photographs taken in the course of field work in Zambia or on holiday in the Florida Keys are worth much more than a thousand words spoken to a painter," he noted, "and much less time is required to take a photograph than to find a similar picture in a library."

Russell visited major collection localities in several parts of the world. In 1974, for instance, he was able to identify New Zealand's first dinosaur bone in the collection made by Joan and Pont Wiffen in the Te Hoe Valley. In 1977 and 1978 he visited Cutler's Tendaguru site, now in Tanzania. The First World War German expeditions alone, he noted, had put twelve times more person-days in the field than Canada's entire effort to date, and attempts to reopen the site would not succeed without earth-moving machinery. Russell's 1989 book on North American dinosaurs, *An Odyssey in Time,* is illustrated by his photographs from such far-flung spots as Kenya, Indonesia, Java, New Caledonia, and Zambia. By the beginning of the 1980s he had visited every continent but Asia, so he was ready when an opportunity came along.

DINOSAURS FOR PEACE

At the Provincial Museum of Alberta in the 1970s such travels were not conceivable — not from lack of imagination or vision, but because of the realities of limited funding and a huge backlog of neglected opportunities in our own backyard. However, barriers were broken by collecting dinosaur footprints next door in B.C., and soon Phil Currie was visiting museums in other countries that held Alberta dinosaur material. When Dinosaur Provincial Park was nominated as a World Heritage Site in 1979, the province was beginning to look for international recognition.

In the red-brick Boardwalk office building in downtown Edmonton in the early 1980s, the staff planning what became the Tyrrell Museum watched the office towers rising around them and dreamed up a new museum. The level of funding and visitor planning studies showed that the new museum clearly would not be just a parochial Alberta facility; it would attract substantial tourism. Those of us concerned with the fossils knew that there were many wonderful stories to be told, and my interpretive planning studies opened many doors to innovative techniques for telling the stories. A key figure was Brian Noble, imaginative communications officer for the project.

Noble had trained as an anthropologist, but was fascinated by dinosaurs, and had worked as a naturalist at Dinosaur Provincial Park before joining the planning team. One day, Noble asked Phil Currie where he would go to hunt dinosaurs if there were no financial or logistical limitations. Without hesitation, Currie answered, "The Gobi Desert." Currie had cut his palaeontological teeth on Roy Chapman Andrew's books about the American Museum's Central Asian expeditions, and was also realizing by now that Canadian and Asian dinosaurs had a lot in common.

Noble did not forget that answer, and when he decided not to move with the museum to Drumheller in 1982, he needed a new project. Inspired by the Russian–American space program organized by Carl Sagan, he felt the same thing might be possible with dinosaurs and anthropology — "a sort of dinosaurs for peace," he said later. In ongoing discussions with Currie, Noble conceived a foundation that would raise the funds to send them both to Mongolia, and called it Ex Terra — from the earth. The idea was ambitious. "Scientists in Asia would be attracted to the project with the opportunity to work with the Canadians in Alberta's extraordinarily rich dinosaur fossil fields, and Canadians would join Mongolian scientists to work sites in the Gobi Desert," explained Noble. Many specimens would be gathered, and "an exhibition could travel the world showing dinosaur relations between two regions of the planet, and other project programs such as films and publications would bring the scientific results and cutting-edge ideas to an international audience."

In 1983, the Canada Council provided an $8700 explorations grant to try to set up a Mongolian–Canadian scientific exchange. Tyrrell Museum Director David Baird was uncooperative, suggesting, "I don't think this is the kind of thing a bunch of amateurs should be doing." But Alberta Culture (the department responsible for the province's museums) provided another $10,000 for the foundation to work on a marketing plan for an exhibit that might result. Noble brought in Glenn Rollans (his collaborator on the book *Badlands*) to approach museums around the world that might want the exhibit, and by the end of 1983 forty had expressed interest.

"DINOSAURS WITH DR. CURRIE"

In the 1920s, Roy Chapman Andrews — a real-life prototype of the movie character Indiana Jones — led a series of expeditions to Central Asia, and his vivid accounts suggest he fought off bandits with one hand while collecting fossils with the other. Mongolia was nominally independent, but both China and Russia laid claim to it. As most of it was in the Gobi Desert, they lacked the motivation to do much about their claims. From his winter quarters in Peking, China, Andrews led scientific teams by motor vehicle, supported by huge camel trains that packed fuel in and packed specimens out. Their finds included many dinosaurs and other fossils, and the New York displays, astute publicity gimmicks, and Andrews' string of books made the expeditions famous.

By the end of the 1920s growing Chinese nationalism made Andrews' work impossible. Then, the Japanese invasion during the Second World War and the Chinese communist revolution closed China and Mongolia from the south and the east, while the growing power of the Soviet Union closed it from the west. Mongolia (formerly Outer Mongolia) was now a state of the U.S.S.R., while China swallowed up Inner Mongolia as one of its autonomous regions. Polish and Russian scientists were able to collect dinosaurs on the Russian side, working with a few Mongolian palaeontologists. China developed its own specialists, who were finding dinosaurs in many parts of China. When Ex Terra was founded, scientists outside the communist bloc had been barred from Mongolia for half a century; the U.S.S.R. was still intact, and China had made only the most limited concessions to Western interest.

Noble approached Moscow, and wrote to the director of the museum in the Mongolian capital, but had no success. Dale Russell of the National Museum was by now an enthusiastic participant, and had suggested that the Canadian Arctic was an important link between North America and Asia. But Russell was also interested in sauropods, which were being found in Xinjiang, a far western province of China. Canada had prior connections with China through Davidson Black (1884–1933), who was the first honorary director of what became the Institute of Vertebrate Paleontology and Paleoanthropology (IVPP) in Beijing (and had scooped Peking Man from under Andrews' nose). Canadian doctor Norman Bethune had helped Chairman Mao's Red Army and was a hero to the Chinese.

In 1985 Alberta businessman Erick Schmidt offered to contact Chinese scientists while on a trade mission in China, and Noble jumped at the opportunity. Schmidt approached Chang Meeman, head of the IVPP. The Chinese had worked unhappily with the Russians, and had been cool to other approaches from France and the U.S. However, France did not have many dinosaurs of its own, and the Americans had not offered the Chinese the opportunity to research in the U.S. A rather tentative letter of agreement was drafted and signed, and the project was — however shakily — under way.

In August Noble flew to meet Chang Meeman at a conference in Tokyo, and later in the year he was invited to Beijing. There, a young graduate student, Yu Chao, told him that "I would very much like to study dinosaurs with Dr. Philip Currie." Obstacles began to fall away. In November 1985 I was present in Edmonton when a Memorandum of Understanding was signed at the Provincial Museum by Sun Ailing, assistant director of the IVPP; Dong Zhiming, China's leading dinosaur scientist; Charles Gruchy of the National Museum of Natural Sciences; and William Byrne of Alberta Culture. "Dinosaurs will bring us closer together, and bring friendship between our two countries," said Sun Ailing, smiling and clapping back as her remarks were applauded. Currie answered questions about the scientific objectives. The exhibit would use Canadian display design and technology, and feature interactive computer systems and computer graphics. Its budget for design and travel would be $20 million. Time lines showed the exhibit development commencing in 1986, and touring from 1989 to 1993.

FROM XINJIANG TO AXEL HEIBERG ISLAND

At the beginning of May 1986, Phil Currie, Dale Russell, and Brian Noble flew to Beijing, and then headed out with Dong Zhiming to Xinjiang, west of Inner Mongolia. There in the Junggar Basin, dinosaurs from every period of the Mesozoic had been found, but many sites remained unexplored. Several promising sites were located for the first full field season the next year, and a camp site was selected in the area known to Dong Zhiming as Dinosaur Valley. Little more than a week later they were back in Beijing to make detailed plans. They talked with Dong Zhiming and his colleague Zhao Xijin, a specialist in sauropods. Currie and Russell bashed out a scientific agreement on Dong Zhiming's typewriter. It included three years in the field and the involvement of twelve Canadians and twenty-two Chinese, and acknowledged that all specimens collected would "remain the property of the country of origin."

The IVPP raises its own funds, and although the car repair shop it runs on the premises might seem a bit odd to Western eyes, its other business of selling eggs from a couple of thousand chickens is really a form of dinosaur ranching. The joint field program provided more income, for the Canadian partners put up two-thirds of the estimated $80,000-a-year cost of the field program.

By June, five Chinese palaeontologists, including Li Rong and Tang Zhilu, were working in Dinosaur Park, with an expense budget of one dollar a day from home. They had never been outside China before, and spoke no English. Peter May, the Tyrrell technician running the museum's field program, helped them to choose groceries paid for by the museum, while they helped him to collect a *Centrosaurus* skull and a fossil log. Tall Tang Zhilu was chief technician at the IVPP, and already had several years of field experience in China. On August 23, three days before they were due to return to China, he spotted and excavated a small piece of bone that proved to be part of a skull. Technician Kevin Aulenback carefully exposed it, and recognized it as a braincase — but of what?

Currie was not immediately available, for he was in the Canadian Arctic. Dale Russell had been working in the north for years, and his team from the National Museum were in a sense hosts of the Tyrrell's Currie and the IVPP's Dong Zhiming and Yu Chao, who had now got his wish to study with Currie. Since the discovery of *Arctosaurus,* a number of vertebrate fossils had been found in the Arctic Islands, but they were all marine reptiles.

Connections between similar Asian and North American dinosaurs might have been through a land bridge to the west (an ancient version of the ice age Bering land bridge), or to the east through Greenland and Spitzbergen. Both were likely, as the north slope of Alaska had yielded bones and teeth of duckbills and horned dinosaurs, tyrannosaurs and troodonts, while Spitzbergen had produced footprints of hadrosaurs and carnivores. Russell had hopes for Axel Heiberg Island, west of Canada's farthest north Ellesmere Island. Both had exposures of the Lower Cretaceous Isaacsen Formation, containing coal seams indicating similar conditions to dinosaur localities in Alberta and British Columbia.

The party ended up on the island without their luggage, but with temporary supplies loaned by the Polar Continental Shelf Project, which provides logistical support for many Canadian researchers in the Arctic. Rifles were carried in the field; the Arctic's top predator is the polar bear, perhaps as dangerous as *Albertosaurus* must have been. In collaboration with University of Saskatchewan palaeobotanist Jim Basinger and other researchers, the party split up to check a number of likely places. Plants, marine reptiles, molluscs, and trace fossils were found, but the only scrap of bone that could have been dinosaurian proved to be a piece of palm tree. "In the end we had nothing," said a disappointed Currie.

At the Dinosaur Park field station, Kevin Aulenback handed Currie the puzzling braincase. "I just went through the roof," Currie remembered. He knew he held in his hand a key piece of the puzzle of the little dinosaur *Troodon*. It was the most complete skull found, and Currie showed it to me soon afterwards. Cradling the fragile bone in his hands, he pointed out the system of air passages that connected the middle ears, a feature also found in birds.

In the first year of the Canada–China Dinosaur Project, three joint expeditions had explored two areas of Canada and one area of China. In the Arctic they had drawn a blank — but even negative evidence was useful. Remote Xinjiang had shown the promise needed to justify a major expedition. And the most startling find had turned up in the well-trodden ground of Dinosaur Park.

"POACHED EGGS IN CONCRETE"

The rib was sticking out of the sandstone cliff when Zhao Xijin spotted it. It might have been an isolated bone, and it was overlain by an estimated 100 tonnes of overburden. In this remote spot in western China, it would have been easy to leave it for better prospects, and the crew thought about it for several days before taking any action. But the Chinese were ready to use dynamite, a technique abandoned in Canada years before. The rib proved to be the only visible part of the first complete skeleton of a *Mamenchisaurus* ever found. The 26-metre (85-foot) sauropod had an astonishing 13-metre (43-foot) neck — "the longest neck in the history of life on earth." The bones were so fragile and the matrix so hard, that Russell compared it to digging a poached egg out of concrete. It took four years to excavate its skeleton completely, and the new species was named *sinocanadorum* in honour of the joint Chinese–Canadian project.

The 1987 expedition to Xinjiang was serious business. There were forty-five in the crew: thirty Chinese and fifteen Canadians, plus a five-person film crew. By now the scientists had developed great respect for each other. "Dong Zhiming is just a superb field man, more concerned with the finding, collecting and description of the specimens," explains Currie on camera. "Dale Russell is . . . the thinker of the three of us, the one that develops the ideas."

The field headquarters was a Calgary-made structure, which had to be erected in 54 degrees Celsius heat; so the team appreciated its generator-powered air conditioners, freezer, and refrigerators. They explored with five Jeep Cherokees, purchased with a grant from the Donner Canadian Foundation.

The *Mamenchisaurus* was not the only find. Among more than two dozen skeletons were (to Currie's delight) a new, almost complete carnivore, later named *Sinraptor dongi* — or "Dong's Chinese carnivore." In September, Emlyn Koster (a sedimentologist who was now director of the Tyrrell Museum) and Chuck Gruchy of the National Museum joined Noble, Russell, Currie, and Zhiming on the return journey. At Bayan Mandahu on the Chinese border with Outer Mongolia, twenty-one skulls and skeletons of the small horned dinosaur *Protoceratops* were found. At Iren Dabasu, one of Andrews' camps, Currie picked up a handful of bones that were very similar to Alberta material. "I could have picked up this same handful in Dinosaur Provincial Park," he commented. Back in Beijing, a press conference attracted twenty-five journalists from most of the world's major news services.

AFTER THE BULLDOZERS

In 1988 new members joined the Canadian team. The Tyrrell Museum fielded invertebrate specialist Paul Johnston, while the Canadian Geological Survey sent Tomasz Jerzykiewicz to work on sedimentology. Jerzykiewicz had a unique background; as a Polish geologist he had participated in the Polish–Mongolian expedition of 1971 to the Mongolian side of the border. He joined the Calgary office of the Survey in 1982, and become an adjunct research scientist of the Tyrrell Museum. CBC radio producer Bill Laws joined them, too, and interviewed Currie and others in the middle of the excavations for the *Ideas* program.

At the Junggar Basin site they continued excavating the metre-(3-foot)-long neck bones of *Mamenchisaurus* and blasted away at the *Sinraptor* quarry until they could extract the plastered skeleton. Another party worked at Bayan Mandahu. "The first day we were out, Kevin [Aulenback] found . . . a nest of eggs within ten minutes of walking away from the truck," reported Currie during the radio interview. IVPP graduate student Zheng Zhong found five *Protoceratops* skeletons facing in the same direction; Currie decided they had been buried while sheltering behind a dune in a sandstorm. A similar situation was presented by five skeletons of baby ankylosaurs — *Pinacosaurus.* The truck left the site carrying several *Protoceratops* skeletons, other dinosaurs, and nineteen mammals.

While the specimens were shipped back to Beijing, the team moved on to Iren Dabasu, on the border. On the first day, Currie "examined peculiar little hills of sand covered with fossil bones." He made inquiries and found that the Chinese and Russians had worked together until diplomatic relations had been severed in 1960. They "had used bulldozers to strip the rock lying over a rich layer of bone . . . in the process some bones had been bulldozed onto the piles." A leg bone turned out to belong to the small bird-like carnivore *Avimimus,* and the small tyrannosaurid *Alectrosaurus* was also found. A Lower Cretaceous site, Abuchaideng, yielded remains of psittacosaurs — "parrot reptiles" — and an unexpected stegosaur that seemed to be the world's latest. Another small skeleton collected at the time was thought to be another *Psittacosaurus,* but when Clayton Kennedy prepared it in Ottawa the following winter, it proved to be a small troodontid, *Sinornithoides.* So much material was shipped back to Beijing that the IVPP ran out of storage space.

MEANWHILE, BACK IN THE TRIASSIC

Ex Terra was also working in Canada, but ironically the most exciting news came from Nova Scotia, resulting from a trickle of small finds made largely by American researchers. During his years at the National Museum, Wann Langston looked again at *Bathygnathus,* the supposed dinosaur of Prince Edward Island, summarizing his research in 1963. Ironically, by this time real dinosaurs had been found in the Maritime provinces. How could dinosaurs possibly hide out in an area that had been studied geologically for well over a century, and adjacent to New England where important dinosaur remains had been known for even longer?

During a long career at Princeton, Donald Baird studied many early Mesozoic creatures and became a leading fossil footprint specialist. He worked north from New England onto the more southern shores of Fundy, but pickings were slim. After more than twenty years, Baird summed up his Nova Scotia collecting with Paul Olsen rather ruefully: "Since 1959, persistent collecting has produced a scrappy herpetofauna from the Wolfville Formation . . . and one local ichnofauna." Although fragmentary, Baird had the oldest ornithischian dinosaur outside Argentina, and (at Paddy's Island near Wolfville) ornithischian tracks (*Atreipus*) and others of a theropod dinosaur.

A bigger discovery came on a rainy day in October 1970. Paul Olsen (at the time Baird's graduate student at Princeton) was in Nova Scotia exploring the rocky shore near Parrsboro, on the north shore of the Minas Basin. He spotted fragments of bone on the beach, apparently washed by the rain from the cliff. That night, he watched while

Donald Baird, and preparator John Horner (who later made his own mark in dinosaur studies in Montana) studied his finds. They pieced together a neck vertebra of a medium-sized prosauropod, a 4-metre (13-foot) ancestor of the great sauropods.

Through the Minas channel from the Bay of Fundy flood the world's highest tides — 15 metres (48 feet), twice a day. The tides continuously scour the red sandstone cliffs — about 200 m.y. old. At first they were regarded as Triassic, but Olsen's patient work has shown them to belong to both sides of the Triassic-Jurassic boundary.

Another of Baird's associates was Parrsboro resident Eldon George. Having injured his right arm in a childhood accident, George was unable to play baseball and turned to geology. In 1948 he opened Canada's first rock shop in Parrsboro, and (encouraged by Baird) began to find bones and footprints. On April 10, 1984, he spotted shallow imprints in sandstone exposed on the shore. "My God, those look like tracks," he thought. No bigger than a penny, they are perfect in detail, with pads and claw marks — the smallest dinosaur tracks known.

Olsen continued to sleuth the basins of the Newark supergroup, but in fifteen seasons he found very little bone. Later, Harvard student Neil Shubin was collecting in Arizona but felt he should look for a promising research project closer to home. He decided to search for an exposure of the Newark supergroup (which extends from Nova Scotia down through New England) that wasn't covered under city lots or forest. Olsen (by now at Columbia University) drew his attention to Fundy.

A first season in 1984 seemed wasted until a jaw of a mammal-like reptile turned up in Shubin's samples, enough encouragement for his work to continue the next year. One day, Olsen and Shubin, working with Hans-Dieter Sues (then of the Smithsonian), were delayed by tides, and Shubin found a white speck in a dark lava. He suddenly realized there were hundreds of fossil fragments like this. "Bones were sticking out all over the place," he rejoiced. "It looked like Rocky Road ice cream." In January 1986 Olsen and Shubin announced the find in the National Geographic Society's press room in Washington, D.C.

Ongoing detailed work produced bones and scutes, teeth and skulls — remains of small creatures that had hidden or died among basalt boulders. Later (working with Robert Grantham of the Nova Scotia Museum), partial skeletons of prosauropods (adult and baby) turned up. With support from the National Geographic Society, three tonnes of rock from this bone-rich bed were crushed and searched. Eventually, ten thousand bones of crocodiles, lizards, sharks, fishes, and large and small dinosaurs were collected.

Now, Nova Scotia is known to have the oldest dinosaurs ever found in Canada, and the tiniest dinosaur footprints known anywhere in the world. It is no wonder these fossils stayed out of sight for so long. Although some are larger, many are very small — the size of a match stick. It can take hundreds of hours of preparation before some specimens can be studied.

Olsen sums up the importance of the region. "This is the only place where you can see the change in the animal assemblage through that critical time when dinosaurs began to rule the earth. You can actually see the opening of the age of dinosaurs right here . . . On the other shore, they didn't rule the earth yet, and here, they did."

BONES ON BYLOT ISLAND

In 1987, Joshua Enookalook walked into the camp with a handful of specimens. They looked like bones, he told his boss, Elliott Burden. Memorial University geologist Burden agreed with his Inuit assistant — they were fossil bones. The place was Bylot Island, close to the northern tip of Baffin Island. It was more than 500 kilometres (310 miles) south-southeast of Axel Heiberg Island, where the Ex Terra team had drawn a blank.

Burden had trained in Calgary and had met Dale Russell in Alberta. With a thirty-minute stopover in the Ottawa airport in the fall, he phoned Russell, who rushed to the airport to look at his bones. The first specimen Russell unwrapped was a toe bone from a duck-billed dinosaur.

In June 1988, Russell and Burden were together on Bylot Island. At Enookalook's site, they quickly found leg bones of a duckbill, a possible bird-mimic, the flightless bird *Hesperornis,* and a small carnivore tooth. Despite the extensive snow cover, a helicopter flight and closer examination of other sites helped to sort out the geology. The Late Cretaceous beds were marine but had been deposited close to the shore. The dinosaurs had died near the shore, and their bones had been broken up by marine reptiles before they were deposited. The material was poor; it was unlikely that there would be skeletons here. But they had located the world's most northerly dinosaur bones! Russell headed for China with the news.

In 1989, the Arctic trip was scheduled for July: there would be a better chance of seeing rock instead of snow. Currie and Russell were accompanied by several colleagues, and Dong Zhiming was with them. There was a lot more bone, much of it from marine reptiles, but some dinosaurian. The party moved on to Ellesmere Island, the most northerly part of Canada. Dale had been told of a fossil skeleton found in an illegal hunting expedition half a century earlier. They found lots of marine reptiles — plesiosaurs and one mosasaur — but no dinosaurs.

"DINOSAURS CRY FOR BEIJING STUDENTS"

On their first night in Beijing, the Canadian advance party for 1989 went walking after dinner. They met a huge crowd of students returning from a demonstration in Tiananmen Square, carrying banners. The growing numbers of students assembling in the square were calling for science and democracy.

Russell was accompanied by Clayton Kennedy, while the Tyrrell Museum was represented by turtle specialist Don Brinkman. Linda Strong-Watson, who had been a technician at the Provincial Museum and then at the Tyrrell, was now science director for Ex Terra. (Currie was in New York examining Andrews' field notes, and Dong Xhiming was in Sweden.) After a week's delay, they headed for Xinjiang, and began work in the field. Disturbing rumours came from Beijing and Russell had a severe heat stroke, but collecting continued. Strong-Watson returned to Beijing, finding the city in chaos, and flew on to Canada.

The rest of the team were at a banquet on June 4 when news came through of serious trouble in Beijing. All internal flights were cancelled; there was nothing to do but carry on with fieldwork. The Chinese members of the team were all from the city and were deeply worried about their families. The party moved back to the Junggar Basin site, where they managed to hear BBC radio broadcasts and learn details about the dreadful confrontation between students, workers, and the army on the night of June 3. Soon a message reached them from the Canadian Department of External Affairs, telling them to leave China as soon as possible. Ironically, international flights were leaving from Beijing, and the Canadians were stuck there for the weekend before they could leave. They learned that someone from the IVPP had sent refreshments to the students in an IVPP van displaying the slogan on the outside that said "Peking Man and Dinosaurs Cry for Beijing Students." Sun Ailing was expecting arrests and reprisals, and the Canadians, heading by taxi towards the square, were met by rifles pointing at the windshield. They escaped on the last Canadian Airlines flight, wondering if they would ever return. They must have reflected that, just as had happened with Andrews' expeditions, Chinese politics had once again got in the way of international science. But although there were disastrous consequences for those in the square, the expeditions were not directly affected.

By August, Strong-Watson was back in Beijing with the Ex Terra administrator and Kevin Taft. One year had been added to the original three-year agreement, but the agreement had been cut short before work could be completed at two important sites. The Chinese scientists were worried that the Canadians would want to pull out, while the Canadians were worried only that the changed political climate would make it impossible to complete the planned joint work. Beijing, however, had changed. There were bullet holes in the buildings, and the sinister presence of soldiers.

"WHAT WERE THE TEETH LIKE?"

In June 1990, Currie, Russell, Strong-Watson, and their colleagues were back at the IVPP in Beijing. As soon as personal questions had been answered, Russell asked about the *Mamenchisaurus.* Work was already in progress at the quarry, and Tang Zhilu had just phoned to say they had followed the neck into the rock, and were finding pieces of the skull.

Since so many dinosaurs lose their skulls before burial, and they are essential for identification, this is always good news. But a *Mamenchisaurus* skull was of particular importance to Russell. Previous partial skeletons had been without skulls, and teeth were critical to compare with other types from North America. This evidence could show how closely Chinese sauropods were related to North American ones. Evidence from other fossils suggested that Asia and North America had been separated during the early Cretaceous, allowing independent evolution from the basic types of animals common in both areas. Russell was anxious to see if this was also true for sauropods, and had waited patiently since the discovery of the *Mamenchisaurus,* while its long neck had been followed into the rock. Three days later, more skull fragments and some teeth were reported. "What were the teeth like?" was Dale's immediate question. He was delighted to learn that the teeth were different from those of its North American relatives.

While other members of the team explored new territory, Currie and Dong Zhiming worked at Bayan Mandahu. Soon they had another group of seven juvenile *Pinacosaurus,* lying parallel, perhaps buried in a sandstorm. Another find had a small dinosaur crouching over a nest full of eggs. The American Museum expeditions had made a similar find from a site not far across the Mongolian border. Since the eggs were thought to be from the horned dinosaur *Protoceratops*, the American Museum's Osborn had reasoned that the dinosaur had been robbing the nest. He called it *Oviraptor philoceratops*, or "horned-dinosaur-loving egg stealer." Another identical find was too much of a coincidence; it suggested to Currie that the animal might have been sitting on its own eggs instead. Back at Erenhot, Don Brinkman found several nests of eggs — perhaps ornithomimid.

On his last day at the site at the end of July, Currie walked back to an old border post, site of an American Museum camp in the 1920s. A broken beer bottle labelled "Made in the U.S.A." reminded him of those explorations, and of his own childhood reading of Andrews' books. Now, he had been a leader in four years of fieldwork in that same desert, producing sixty tonnes of fossils, and laying the foundation for future cooperation with Asian colleagues.

Meanwhile, the world had changed. The Berlin Wall came down in 1989, and tentative discussions between a newly independent Mongolia and the American Museum had opened the doors on which Brian Noble had knocked in vain in 1983. In June 1990, the first team from the American Museum since Andrews' day began work across the border, in the Mongolian/Gobi. Soon, the American and Canadian scientists would be able to share their expertise with each other as well as the Asian scientists, as another flood of new material came from Mongolia.

The Sino–Canadian Dinosaur Project (as it was now being called) was important for all the parties. Ex Terra managed to pull together funding on a scale that was new to Canadian palaeontology. It brought on board important sponsors like Canadian Airlines (who facilitated transportation of scientists) and made major financial contributions to the overseas fieldwork — $100,000 a year during each year of the project.

For Chinese and Canadian scientists alike it built relationships that have continued beyond the project. For the Chinese it provided training for their students not available at home, where an entire generation of academics-in-training had been killed, imprisoned, or driven abroad during the Cultural Revolution. For the Canadians it provided access to a region of supreme importance in the understanding of Canadian dinosaurs, and rejuvenation of their own quest for understanding. Dave Eberth of the Tyrrell Museum recollected that "When I was at graduate school, it was pretty much taken for granted that all the major discoveries had been made. . . . Then, when China opened up to me, it was suddenly all there to be done. . . . It was just like Alberta and Montana must have been to paleontologists a hundred years ago."

In 1991, the last official fieldwork for the Canada–China project took place in Alberta, when Chinese scientists joined Tyrrell staff in Dinosaur Park, and at a new footprint site near Grande Cache. By this time, Ex Terra's emphasis had shifted to planning the long-anticipated exhibit.

THE GREATEST SHOW UNEARTHED

As the Chinese fieldwork ended in 1990, Canada's first international dinosaur exhibit was opening its doors in Chiba, Japan. Sponsored by Alberta Tourism in collaboration with Hitachi Corporation and other Japanese companies, Hitachi Dinoventure brought fourteen full-size dinosaur mounts and thirty-three other specimens across the Pacific Ocean. Highlights included a skull of *T. rex* being cleaned from the rock, and Russell's dinosauroid model. The $1 million exhibit mounted by the Tyrrell Museum ran from July 7 to September 2, and was projected to draw two million visitors. It served as a pilot for the more ambitious exhibit being planned for the Canada–China project. Although magazine articles and television spots shared some information about the expeditions, their real success became apparent to the public only with the opening of the Dinosaur World Tour exhibit.

In Edmonton, Ex Terra had been gearing up to develop an exhibit based on the data and material gathered in its field research. As the organization changed gear, new funding sources were needed, and critical questions had to be asked. In 1988 Ex Terra alternatives varying from 750 to 1860 square metres (8000 to 20,000 square feet) were being considered. A dozen cities would get the exhibition, opening in 1991, travelling for four years, and being seen by twelve million visitors. By 1990 the foundation was contemplating an exhibit of 3200 square metres (40,000 square feet). As the stakes got higher, board and staff politics got more complex. Eventually, Brian Noble, who had the original idea and got Ex Terra under way, left the project.

Jim Ebbels had been brought in as Director for Exhibits. His background was in architecture, and he had worked on interpretive centres. "I was hired to get the exhibit designed and on the road," he told me. "With the changing of the guard in '91, I became CEO." On the growing staff, two had a strong dinosaur background. Linda Strong-Watson was now Director of Science and Public Programs, while former Dinosaur Park naturalist John Acorn became science consultant for the exhibit. The main design was undertaken by Vancouver-based Aldrich

The Greatest Show Unearthed.

Pears, and partners included the National Film Board and publishers Macfarlane Walter & Ross, and Greey de Pencier. Dozens of smaller companies and individuals worked on different segments of the project.

Projected exhibit opening targets of 1989, 1991, and 1992 went by as the small team raised money, canvassed for support, and brought together specimens, information, and resources in their Edmonton headquarters. The biggest skeleton, of *Mamenchisaurus*, was so large that it had to be constructed in the parking lot.

At last the exhibit opened on May 14, 1993, in a huge white tent near Edmonton's convention centre, across from the Muttart Conservatory. The main tent was 2300 square metres, (nearly 25,000 square feet) and included thirty dinosaur skeletons (real or casts) — a mere ten percent of the bones collected by the expeditions. The dinosaurs were supported by many other exhibits, and the show was optimistically entitled the "Dinosaur World Tour — the Greatest Show Unearthed."

The public entered the exhibit through Albertosaurus Avenue, containing Brian Cooley's life-size *Albertosaurus* model flanked by a collection of dinosaur illustrations in many forms — in ads, models and kitsch. Meat Eater Theatre showed clips from monster movies, then attendees were confronted by a stunning cast of the Alberta *T. rex* called "Black Beauty." The briefing theatre used video to introduce the expeditions (using a terminally cute baby dinosaur named "Gobi"). In the dramatically lit main exhibit, mock-ups of a desert camp and a simulated *Pinacosaurus* quarry showed the field side of the expeditions. Alberta exhibits included embryonic *Hypacrosaurus* from Devil's Coulee, *Pachyrhinosaurus* skeletons from a Grande Prairie bone bed, and Grand Cache tracks. Chinese dinosaurs included skeletons of the carnivores *Sinraptor* and *Monolophosaurus,* the ceratopsian *Psittacosaurus,* the new segnosaur *Alxasaurus,* and *Mamenchisaurus.* The *Sinornithoides* skeleton, still in its plaster jacket, was supported by models of the little bird-like dinosaur. The viewer exited past a mock-up of decaying stegosaur, being scavenged by troodonts, pterosaurs, and beetles, and arrived in the dinosaur store, heavily stocked.

Kids (and uninhibited adults) could observe fossil preparation taking place, participate in a simulated dinosaur dig, cast dinosaur tracks, and — at the Discovery Stage — get a guided tour of another *T. rex* skeleton. Paintings were by leading Canadian dinosaur artists, and a pioneer computer graphic showed a hunting *T. rex.* The exhibit was mounted and run with the help of more than 1000 volunteers, and in Edmonton attracted 225,000 visitors and brought some $15 million into the local economy.

"It's one thing to build an exhibit, it's quite another to design one and market one as a tour," said Jim Ebbels in an interview. The exhibit moved to Toronto for five months, and then went to Osaka, Japan, in 1994. From there it went to Singapore and Vancouver in 1995. It then moved to Sydney, Australia, by which time about 2.5 million people had seen it.

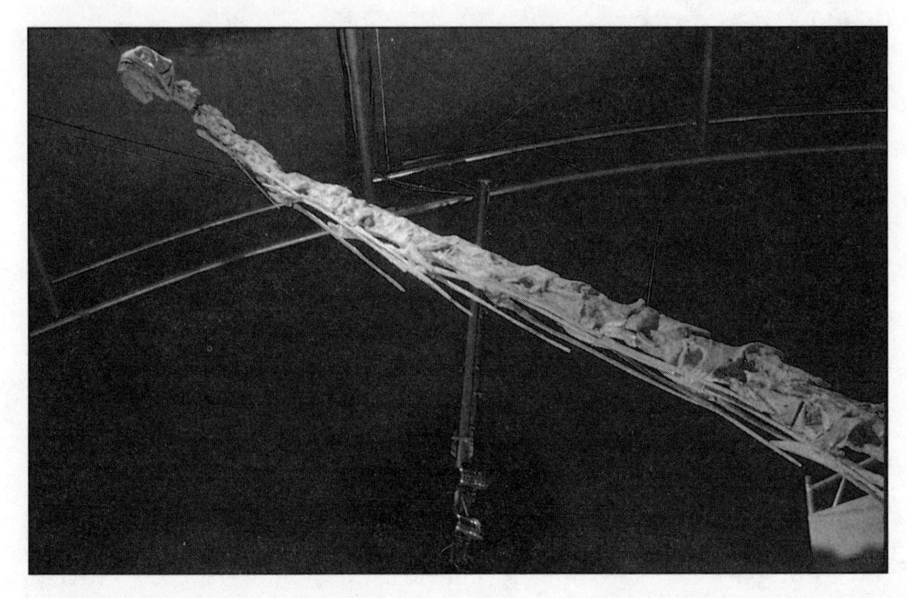

Mamenchisaurus *sticks its neck out.*

A MUCH GREATER ACCOMPLISHMENT

As the fieldwork wound down, institutions participating in the Chinese projects shifted emphasis to preparation and research. In 1993, Currie summed up. "Ultimately," he said, "millions of people will experience this public tribute to the Dinosaur Project. But for the palaeontologists and geologists . . . this special issue represents a much greater accomplishment." He was introducing a 275-page special edition of the *Canadian Journal of Earth Sciences,* which contained many papers arising out of the Canada–China Dinosaur Project. Many were jointly authored by Canadian and Chinese scientists working together. A small proportion of the discoveries produced enough material to describe eight new genera and eleven new species of turtles and dinosaurs. Other papers described the sediments in which they were found and analyzed the palaeogeographical evidence derived from the expeditions.

Papers were included on the sauropod *Mamenchisaurus,* the carnivore *Sinraptor,* the crested *Monolophosaurus,* the smaller *Alxasaurus,* and a new species of stegosaur, *Wuerhosaurus ordosensis.* Other significant dinosaur specimen descriptions included embryonic primitive horned dinosaurs, a detailed description of a *Saurornithoides* hind leg, and of the *Troodon* skull found in Dinosaur Provincial Park. Among turtles and lizards was a description of a partial bird skeleton. A second volume of reports published in 1996 added another 137 pages of data, including two important dinosaur papers. One paper concluded that the *Oviraptor* skeleton at Bayan Mandahu was probably sitting on its own eggs — and perhaps even laying them at the time of its death. The second describes two new species of *Psittacosaurus,* noting that this small genus was now one of only four dinosaurs known from more than a hundred skeletons or partial skeletons.

DISCOVERY OF THE CENTURY?

"Take a bird and pluck the feathers off and you have a dinosaur," says Phil Currie. The idea that birds and small carnivorous dinosaurs are closely related has been around since Huxley suggested it in 1868. For years, the absence of the furcula (wishbone) from the skeletons of the carnivorous dinosaurs that seemed closest to birds ruled out a close relationship. In birds this bone is fused from two clavicles (collarbones) that help the body support the stress of flapping wings. Then, specimens found in Mongolia showed that this bone was present in some dinosaurs.

The skull of *Troodon* found by Tang Zhilu in Dinosaur Provincial Park showed features normally crushed out of recognition. CAT scans showed the presence of air passages connecting the middle ear over the skull (perhaps increasing sensitivity to low sounds). The only other animals with these passages are crocodiles — and birds. The list of similarities between birds and dinosaurs was growing. "Dromaeosaurs and birds have 80 shared derived characters in common," comments Hans-Dieter Sues, "with scores and scores of features in different parts of the skeleton."

Theoretical discussions revolved around the origin of feathers, which for centuries had been regarded as the one unique characteristic of modern birds. Although rarely preserved, they had been found on a number of fossil birds, including *Archaeopteryx* from way back in the Jurassic — the middle of dinosaur time. Had the feathers first evolved for flight or for some other purpose? Theorists argued that feathers were obviously modified reptilian scales, which would not be useful for flight until they were well developed, so that they must have had another function first. With the renewal of the hot-blooded debate, the question had a new importance. "Small predatory dinosaurs . . . may have evolved insulation to maintain a constant high temperature. . . . Perhaps feathers initially evolved in dinosaurs as insulation and were secondarily adapted for flight in birds," explained Currie in his 1991 book, *The Flying Dinosaurs,* adding regretfully, "We may never know."

In the same book Currie discusses a newly discovered feathered fossil from Liaoning in China — an unnamed bird. From the same area in 1996 came other creatures that had "feather-like structures covering the head, trunk, tail, arms, and legs," that were "simpler than true feathers," and "lack the aerodynamic quality of avian feathers." It was regarded as a close relative of a little dinosaur from Europe, and later named *Sinosauropteryx* ("Chinese feathered reptile"). Study of other features of this creature suggested the idea that feathers may have evolved in juvenile dinosaurs, being generally lost as they grew into adults, and explaining why they had not been found previously. Birds could have evolved by the retention of these juvenile features into adult life, a phenomenon familiar in other groups of animals. (An often-quoted example is the human species, as our large, smooth skulls resemble the juveniles of our ancestors more than the heavily ridged skulls of adults.)

In 1998 Currie with two Chinese scientists, Ji Quiang and Ji Shu-An from the National Geological Museum, and Mark Norell of the American Museum of Natural History, described two remarkable new Chinese dinosaurs with feathers. Currie remembers his reaction when he first saw one of them in Beijing in 1997: "My jaw dropped. Not only was it a complete specimen the size of turkey, but it also had structures along the back that were feather like." Based on careful study of the original specimens by Currie, the new finds were named *Protarchaeopteryx* and *Caudipteryx.* The specimens were adults, and were identified as coelurosaurs, not birds. "The presence of . . . feathers on non-avian theropods," concludes the article firmly, "provides unambiguous evidence supporting the theory that birds are the direct descendants of theropod dinosaurs."

It was this research that earned Currie his place on the cover of *Time.* The popular story underlines the key messages of the technical paper. "It is one of the most exciting discoveries of the century, if not the discovery of the century." Less than a decade after regretting that "we may never know," Currie does know: "It's not that birds are descended from dinosaurs. It's that birds *are* dinosaurs."

12

"HOW CLEVER WE ARE"

"I'm the dinosaur. I dream of what I would do, how I would eat, how I would feed my baby and how I would protect myself."

ELY KISH

LOST IN THE BADLANDS

"All around us . . . were buttes, hoodoos and eroded gullies and hills, in which time itself is a visible component . . . you are actually looking at time. . . . In the badlands of the Red Deer River are the ancient ruins of earth itself. And they are very beautiful." Poet Al Purdy wrote of the inspiration behind his poem "Lost in the Badlands" (from which I took the title of this book). Like a palaeontologist, he "left prescribed paths" to explore, but his quest was not the palaeontologist's search for bones and facts, but the search for insight and beauty. His product was not a scientific paper or a museum exhibit, but a poem. Dinosaurs have more than scientific significance to us, and it is intriguing to explore the impact of dinosaurs and their discoverers — particularly Canadian discoveries — on our culture.

Canadians know they are affected on a unique scale by the country next door — our usual metaphor is the elephant and the mouse, but perhaps in this context the *Seismosaurus* and the *Troodon* would be more appropriate. Scientifically, Canadian dinosaurs are not as well known — even in Canada — as American ones. When they are, it is often because they also occur in the U.S., or because specimens were collected by Barnum Brown or Charles Sternberg and are now on display in American museums. Major museums in American cities produce more publications and provide a ready resource for publishers, writers, and photographers based there. Their publications are then distributed in Canada, often more widely than those originating from Canadian publishers.

From that distance, Canadian stories can be invisible. Two books published in the last decade by the American Museum illustrate this. "Barnum Brown . . . floated a raft down the rugged Red Deer River in Alberta, becoming the first paleontologist to explore the area's extraordinarily rich Late Cretaceous bone beds," says one author, sweeping aside forty years of Canadian discovery and Lawrence Lambe's early achievements. "We were the first team of paleontologists from the West to explore the Gobi Desert since the 1920s . . ." says another, of the 1990 expedition that started only after four years of Sino-Canadian work in the Gobi. American achievements are enormous, and they have no need to belittle those of other countries. Yet here is another area in which our attention is directed away from our own resources, in which our stories are being replaced by those of others.

As Canada has a branch-plant economy, with all media dominated by the products of Hollywood and the American networks, our culture is swamped by American publishers and performers. Yet in many ways Canada has a distinctive culture that reflects our historical and creative differences. Knowing that dinosaur science is dominated by Americans, we must ask: How far is this true of dinosaur arts?

Certainly, Canadian kids generally know far more about dinosaurs from the U.S. than they do about Canadian ones. They get most of their information (and misinformation) about dinosaurs from the U.S., and sit down to watch *The Flintstones* and *Jurassic Park*, just like kids elsewhere in the world. Do they have the opportunity to find inspiration in the work of Canadian palaeontologists? As Canadian dinosaur research has emerged on the world stage, is this reflected in other products of our culture?

In looking at the impact of dinosaurs — particularly Canadian dinosaurs — on Canadian culture, I am not suggesting that an artist — Canadian or not — does not have freedom to use any topic and region, or to be creative with the facts. However, if Canadian creators chose to write only about other countries, who would celebrate Canadianness?

Dinosaurs have in fact been an inspiration in virtually every aspect of culture practised in Canada — even briefly in music, in Moe Koffman's "Dinosaurus" from *Museum Pieces* (1977). Here are a few highlights.

SNAKE PEOPLE

The First Nations were aware of dinosaurs in Canada before anyone else. As scientific knowledge has grown, their cultures have also developed and assimilated new knowledge. In this century, aboriginal authors increasingly become the interpreters of their own culture to a wider audience, and some of them incorporate dinosaurs into their world view.

Percy Bullchild is South Peigan, a group related to the Alberta Peigans who live in an area that is now in the United States. Though he does not mention the story directly, Bullchild is presumably heir to the traditions of those who knew the buffalo's grandfather. In his wonderful synthesis of Peigan belief, *The Sun Came Down: The History of the Earth as My Blackfeet Elders Told It,* Bullchild describes the time when the Snake People — the first animals Creator Sun made — abounded on the body of Mother Earth. "From their wickedness, these snakes had become many — they were in many forms because of their crossbreeding. Some were beginning to have legs, but they still looked like a snake. And because of no discipline or not wanting to listen [to Creator Sun's rules], many of them became overgrown. Big, big in their form. Tall and long. The life of reptiles, dinosaurs."

There came to be too many, and Creator Sun made it rain "on and on, with no letup." In an apocalyptic vision resembling the biblical flood, Bullchild vividly describes the end of the dinosaurs, some "sinking down into the soft mud-mire," while others were "taking refuge on these hard-surfaced places . . ." until "these places would tip over, dumping these heavy creatures." By the time the rain stopped "there was no heavy reptiles left anyplace." Bullchild explains the layers of the earth, with which "Those creatures of that prehistoric time were covered so fast . . . that they are found, in these days, intact."

Dinosaurs also find their place in First Nations traditions outside areas where their fossils are found. George Blondin presents stories of the Dene (who live in the northern forests of western Canada) in his book *Yamoria: The Lawmaker.* He explains how "our people used medicine power to find answers about our early existence," but that the knowledge thus gained was secret, for "their medicine power did not allow them to talk of what they know." Nevertheless, he lists dinosaurs among the "giant animals that once roamed the earth," which were destroyed by Yamoria, the great medicine man — perhaps a god — who once lived among them.

THE TANNINIM OF OLD

"Palaeontology and the doctrine of evolution are so interwoven that they cannot be divorced," said Parks in 1935. But Christians who interpret the Bible literally have to find other explanations of scientific discoveries. In Alberta, I had friends who told me firmly that dinosaurs had not really existed, but that God had created dinosaur fossils to comfort humanity by making it seem that they had a remote past.

After Charles Darwin published a credible mechanism for evolution in 1859, scientists and theologians (then often the same people) had to explain the apparent conflict between fact and faith. In other countries, public controversy produced such notable debates as T.H. Huxley's encounter with Bishop Wilberforce in England (1860), and the 1925 Scopes Trial in Tennessee. In Canada the debate generally has been less confrontational.

A leading international figure of this debate was McGill's principal, John William Dawson (1820–1899). A dinosaur collector himself, and father of the discoverer of Canada's dinosaurs, Dawson was also an earnest Presbyterian who could not accept Darwin's views on evolution. He produced many popular works attempting to reconcile the geological record with the biblical story. One example was *Achaia: or Studies of the Cosmogeny and Natural History of the Hebrew Scriptures* (1860). Public interest in evolution made Dawson an international bestseller — for instance, *The Story of Earth and Man* (1872) went through eleven official and several pirated editions. Dawson discusses dinosaurs as monsters mentioned in the Old Testament: "the noblest of the Tanninim of old."

Dawson felt that he followed "a quiet middle course, which, however unattractive to the sensation-loving public, is most likely in the end to be correct," but he became increasingly isolated. As a natural scientist who did not accept evolution, in 1878 he was offered a chair at Princeton (then a Presbyterian college). Dawson refused this, and remained an honoured (but ignored) pillar of Canadian science until his death.

Twentieth-century fundamentalists (including Lutheran, Baptist, and Catholic) succeeded in restricting the teaching of evolution in schools in three U.S. states in the 1920s. One leader in this effort was George McCready Price (1870–1963), who was born in rural New Brunswick. A Seventh-Day Adventist, he trained as a teacher with some elementary science courses from one of J.W. Dawson's students. In 1902 his *Outlines of Modern Christianity and Modern Science* invoked Noah's flood to explain fossils. The sequence of life shown by the rocks is explained by their reactions to the flood, the larger animals fleeing to the hilltops and "not immediately covered with earthy deposits." Price moved to New York and then California. A second book, *Illogical Geology* (1906), was republished in 1913 as *The Fundamentals of Geology*, in which Price considered he had dealt "beautifully with every major problem in the supposed conflict between modern science and modern Christianity."

Canada featured in the life of American surgeon Arthur I. Brown (1875–1947), who practised in Vancouver in 1913, wrote antievolutionary pamphlets, and in 1925 gave up his practice to travel as a lecturer, continuing until his death in an automobile accident. The English-based Evolution Protest Movement also developed a chapter in British Columbia with about 120 members in the early 1970s. A later independent Creation Science Association was praised as having been "more active and productive in Canada than in any country except the United States."

Alberta's Social Credit party leaders, William "Bible Bill" Aberhart and Ernest Manning, were premiers from 1935 to 1971. Both were preachers who built their political base through Calgary's Prophetic Bible Institute and its radio programs. In these broadcasts, Aberhart is reputed to have accused palaeontologists of going off into a secret place in the badlands to manufacture dinosaurs from plaster of Paris (presumably as a test of faith for sincere Christians). Charlie Sternberg's wry comment was: "How clever we are." Although there is no proof, I suspect that Alberta's rich dinosaur resource was one of the reasons for the delayed development of scientific museums in the province; believers in creation would not be interested in founding organizations that would be likely to teach evolution. Although oil wealth was flooding into government coffers from 1947, the provincial museum was not developed until Canada's centennial year of 1967.

When I found myself responsible for a provincial palaeontology program I watched with interest for public reaction, but except for occasional reported comments in the dinosaur gallery that "they're all made of plaster," there was no direct response. However, when at one point the Alberta government passed a resolution to give equal time to creation in education I did wonder if I might find myself the central figure in a Scopes-type trial. However, no attempt was made to dictate the contents of the museum galleries.

Building the Tyrrell Museum a mere 60 kilometres (37 miles) from the Bible Institute in Three Hills did raise some fundamentalist response. A group of ministers of different denominations approached senior staff and indicated their intention to oppose the presentation of evolution in the museum. They were tactfully asked to produce a definitive statement of their views for the staff to consider, but sectarian differences perhaps intervened. Once the museum opened, the Institute asked permission to bring tour groups; apparently to teach them techniques to counter the teachings of the museum.

A YUKON CERATOSAURUS?

Less-structured beliefs are part of folklore. The fascinating field of cryptozoology, which depending on the researcher's point of view is either folklore or future science, has recorded a supposed dinosaur sighting in the Yukon! The observations were reported in 1903, when a San Francisco banker and a gold prospector were hunting moose a hundred miles east of Dawson City. "In the snow was the clear impression of an enormous body." Later, an augmented party sighted a monster "black in colour, at least 50 ft long" and estimated to weigh 40 tonnes. A horn on its nose led the priest in the party to identify it as *Ceratosaurus.* As if encouraged by the positive identification, the monster appeared again to the priest in 1908, but has not been reported since.

Laughable as this may be, much equivalent nonsense survives in our everyday discussion. Half-remembered, half-forgotten ideas from school and the confusion of fact and fantasy in the media linger in everyone's consciousness. One often hears, and sometimes reads, about dinosaurs becoming extinct by being frozen in the ice age (losing 63 million years along the way), or that dinosaur bodies are the source of the oil deposits (after all, they both occur in Alberta). According to a poll quoted by Carl Sagan, sixty-three percent of American adults are unaware that the last dinosaur died before the first human arose, and Canadians seem to have the same ideas (after all, thousands of episodes of *The Flintstones* can't be wrong).

"Old fossils join older fossils in Museum," read a headline when the Canadian Senate moved into the Victoria Memorial Museum. Dinosaurs perform another function in our discourse, particularly in headlines, speeches, and political cartoons. They serve as a clichéd metaphor for a loosely associated set of ideas about size, unsuitability for modern conditions, and inability to adapt to changing circumstances. The metaphor itself "fossilizes" an earlier scientific view of dinosaurs, for all of these ideas date from half a century ago, and have all been abandoned by scientists. We now know that dinosaurs were prominent on the earth for at least 160 million years (all humans have managed so far is a mere couple of million years), and evolved quickly to deal with changing circumstances, with the largest and most spectacular types finally (perhaps) only becoming extinct through a cosmic accident, while their apparent descendants (in the shape of birds) survived another sixty million years or so and are still around in great numbers of species and individuals.

It's time to stop stereotyping the dinosaurs — their use as a metaphor should be a strong compliment, and we need to find a new image for the large, inefficient, and out-of-date. Meanwhile, it is gratifying to see new research finding its way into our imagery. A June 1998 letter to the *Globe and Mail* picks up a news story about bird-dinosaur relations that suggests that raptors resemble "large surly chickens. Anyone who's been to a basketball game in Toronto," (home of the National Basketball Association Raptors) continues the correspondent, "could have told them that."

ENTIRELY NEW SCENES

In J.W. Dawson, Canada had one of the first popular geology writers. Although Dawson only mentions his own visit to the badlands in his 1901 autobiography, *Fifty Years of Work in Canada*, his popular books talk about dinosaurs from Britain and the United States, where more complete specimens were then available. Thomas Chesmer Weston's *Reminiscences Among the Rocks* (1899) was the first book to give an entertaining account of dinosaur hunting in the west.

"I . . . will take my readers to entirely new scenes — to the richest Cretaceous fossil field in the world," said Charles Sternberg in his 1917 book, *Hunting Dinosaurs in the Bad Lands of the Red Deer River, Alberta, Canada.* Sternberg's readers were largely in the U.S., for he produced five hundred copies of the first popular book substantially about Canadian dinosaurs in his home town of Lawrence, Kansas. (He produced a new edition in 1932 when he was in his eighties, and I edited a new edition for NeWest Press in 1985 — the first time his book was distributed in Canada.)

Although there were many government reports and technical papers, only the occasional magazine article and museum pamphlet presented Canadian dinosaurs to the public until Dale Russell's *A Vanished World* appeared sixty years after Sternberg's first publication. Splendidly illustrated with Suzanne Swibold's photographs and Eleanor (Ely) Kish's paintings, this is one of the finest evocations of the dinosaurs in their environment.

In the 1980s, the development of the Tyrrell Museum provided a new audience for popular works. Renie Gross's *Dinosaur Country* (1985) brought landscape, dinosaurs, and the story of their discovery together in an entertaining way. Gordon Reid produced the first substantial guide to *Dinosaur Provincial Park* (1986), and Ron Stewart's *Dinosaurs of the West* includes some Canadian content in a more broadly focused work.

Richer content came with Dale Russell's *An Odyssey in Time: The Dinosaurs of North America* (1989), which featured Canadian dinosaurs extensively in a wider canvas. In the early 1990s, Tyrrell staff began to produce popular books. Monty Reid's *The Last Great Dinosaurs* (1990) describes and illustrates the Late Cretaceous dinosaurs of the west, while Phil Currie includes many Canadian dinosaurs and their relatives in *The Flying Dinosaurs* (1991), a pioneer book on the changing views of dinosaur-bird relationships. These (and Currie's 1998 *The Newest and Coolest Dinosaurs*) are often beautifully illustrated by Jan Sovak. Wayne Grady's *The Dinosaur Project* (1993) features the work of Tyrrell staff and others in the Canada–Chinese expeditions, while *The Land Before Us,* (1994) by Andrew Nikiforuk and a number of staff authors ranges more widely, providing a background to all of the Tyrrell exhibits. Recent Nova Scotia discoveries are featured in Harry Thurston's *Dawning of the Dinosaurs* (1994).

After Barnum Brown's expeditions into Alberta, Canadian dinosaurs were sometimes featured in American books, particularly when they had been collected by American museums and were ready at hand. As Canadian research has emerged onto the international scene, Canadian palaeontologists are more extensively featured in books produced elsewhere. Thus, Don Lessem's *Kings of Creation* (1992) starts in Dinosaur Park, and Louie Psihoyos' and John Knoebber's *Hunting Dinosaurs* (1994) also features Phil Currie and Dale Russell. Canadian writers also are producing popular works about dinosaurs on the wider scene. My *Dinosaur Hunters* (1993) is a worldwide history of dinosaur hunting, and Phil Currie is one of the editors of a spectacular new *Encyclopedia of Dinosaurs* (1997).

NOTES ON THE MESOZOIC

Sternberg's *Hunting Dinosaurs* is a landmark of writing about Canadian dinosaurs, for it introduces not only the exploration account that is its main focus, but also pioneers fictional and poetic treatments. Within three chapters of the book, Sternberg vividly but oddly incorporates a fictional time travel story. His dinosaur facts are good, and his interpretations are up to date for the time, even if the human parts of the story read strangely by modern standards. "Evading the snares of the gothic," comments poet Jon Whyte, "Sternberg presents with poetic clarity an almost Edenic vision of a time and a place where his imagination has often transported him in his work."

Sternberg is soon followed by a writer with a greater reputation. New Brunswick's Charles G.D. Roberts, noted as the inventor (along with Ernest Thompson Seton) of the realistic animal story, also may be the first writer to recreate the fictional world of the dinosaurs without a human participant in the opening chapter of *In the Morning of Time* (1919). Unfortunately, Roberts also makes a strange mess of his facts, even by the knowledge of his own day. The book was started in Munich, after Roberts had visited Hagenbeck's zoo and museum in Hamburg. His first chapter, "The World without Man," features *Diplodocus*, an *Iguanodon*, an unspecified carnivore, and a *Stegosaurus* — already an anachronistic mix of creatures from Jurassic and early Cretaceous periods. The anachronisms get worse in the next chapter, as we find a Cretaceous *Triceratops* in battle with *Dinoceras*, an Eocene uintathere, watched from the trees by a Jurassic *Archaeopteryx* and a Pleistocene ape man. However, Roberts' descriptions are vivid and full of action, focusing on the confrontational situations that are favourites in contemporary pictorial reconstructions.

Writer Robert Kroetsch, born near the badlands in Heisler, Alberta, seems to be the first to introduce the life of dinosaur hunters into Canadian fiction in his novel *Badlands* (1975). Kroetsch travelled down the Red Deer River by boat in 1972 to do his research, and I first met him then in a Drumheller pub while I was working on a dig. The book features William Dawe (a dinosaur hunter whose humpback owes something to Dawson, while his Lutheran obsessions and limp are perhaps derived from Charles Sternberg). Dawe travels down river with his eccentric party in 1916, excavating a *Chasmosaurus* and a *Gorgosaurus*. Dawe refers to the earlier travels of Brown and Sternberg, and to researchers Cope, Marsh, and Lambe. The story evokes the lives and work of the collectors, and briefly but vividly the lost lives of dinosaurs ("the huge and bone-cracking jaws finding at last the solid-crested skull"), as Dawe struggles to bring them back to life ("some horny old dinosaur booted in the ribs like a sick cow and told to stand up again"). Dawe and his companions are looking for the specimen that will immortalize them — a sort of dinosaurian *Moby Dick*. But the book is only partly about dinosaurs, and Kroetsch adds rich characters and details — a cabin of bones, a tornado — for his own, more complex, purposes.

Margaret Atwood features the museum side of dinosaur work in her *Life Before Man* (1979). In a transcribed discussion with Alan

Twigg, Atwood explains that the novel was originally going to be "called *Notes on the Mezaoic* [sic] . . . But the title was changed because everybody said 'Notes on the What?'" (Atwood meant, and presumably said, "Mesozoic.") She explains that the novel was "set in the Royal Ontario Museum. And why Lesje is a paleontologist who studies dinosaurs," because (in Twigg's words) "it's the first Canadian novel . . . that seriously conveys an awareness that the human race can become extinct."

Lesje seems to be some sort of curatorial assistant in palaeontology (as she catalogues, edits, and interprets), yet — years after the Provincial Museum and others were employing women in the field — has only ever been on one dig. Lesje's favourite dinosaur is *Deinonychus,* while Canadian dinosaurs are mentioned only in passing because they are on display. Like Sternberg, Lesje frequently imagines herself back in the Mesozoic — though only once in a Canadian setting, where she watches "an ornithomimus, large eyed, bird-like, run through the scrub, chasing a small protomammal." Although she is "monomaniacal about her subject," when in the gallery with children who think the Ice Age is the cause of dinosaur extinction, she does does not correct their error. In her fossil fantasies she "allows herself to violate shamelessly whatever version of prehistoric reality she chooses."

In *The Reconstruction* (1996) the principal character of Vancouver author Claudia Caspar also works for a museum, as a model maker sculpting the early hominid Lucy. The environment she occupies includes "The dark-brown tibia of some lumbering dinosaur whose pinhead bobbed at the end of a neck long and muscular as a python. . . . like a beautiful hallucinogenic mystery . . ." But her dinosaurs, too, are not Canadian, and have little to do with the story.

Canadian dinosaurs turn up in some speculative fiction, but only Harry Harrison, in his *West of Eden* (1984) and its two sequels — inspired by Dale Russell's dinosauroid — seems to be strongly influenced by Canadian ideas. Speculative fiction is now emerging strongly in Canadian writing, but only Robert Sawyer seems to be taking a serious interest in dinosaurs. In a short story, "Just Like Old Times" (1993), Sawyer envisages a future in which a serial killer could be sentenced (with the aid of the Tyrrell Museum) to the body of a *T. rex*, allowing it to hunt primitive primates in revenge. His trilogy of *Far Seer* (1992), *The Fossil Hunter* (1993), and *Foreigner* (1994) features a fantastic world in which the characters are human-acting dinosaurs, themselves conducting geological surveys and finding fossils.

TRUMPETING AT THE SKY

Canadian poets seem to have been moved more directly by dinosaurs. Sternberg's scattered, pedestrian verses have priority, if not inspiration: ". . . . His limbs are armed with claws so great;/ His jaws are filled with horrid teeth./ Alas, I fear our Saurian's fate —/ He's simply dallying with death . . ."

Al Purdy's 1984 collection, *Piling Blood,* includes several dinosaur poems. "Lost in the Badlands" is the longest and most interesting, but "In the Early Cretaceous" gives a vivid image of "a comic-looking duck-billed dinosaur/ might have lifted his head/ with mouth full of dripping herbage/ and muttered Great Scott . . ." "Museum Piece" celebrates museum display, where "dinosaurs they soar/ to fifteen twenty thirty metres . . . and The mind shuffles its feet to think/ of that time: — when diplodocus tyrannosaurus/ and the like trumpeted at the sky/ 65 million years ago . . ." In an article, Purdy talks of places that have shaped his vision, and refers to visits to Dinosaur Park and Drumheller Museum in the 1970s. He gave the name "Albert" to one of the specimens exhibited in Dinosaur Park, a "duckbill dinosaur, with a bun of flesh and bone on top of his (her?) head, like a prehistoric hairdo," and explained its injured tail by an attack in which "*Gorgosaurus* had seized Albert's tail in teeth like piano keys and bitten down hard."

Another poet, Monty Reid, has spent a number of years as a senior staff member of the Tyrrell Museum, and so has a special relationship with dinosaurs. In *The Alternate Guide* (1984) they appear conventionally as an image for "old combines" littering the landscape, "the perplexed dinosaurs who could not adapt." Later and more vigorous poems in *These Lawns* (1990) explore the nature of fossil bone in "Specimen," and Peace River's dinosaur footprints ("movement arrested momentarily") in "Working with Shale."

DINOSAURLAND

"This dinosaur graveyard . . . has no reality in books," says Al Purdy in "Lost in the Badlands." It is no accident that Canadian children often know little about their own dinosaurs, for Canadian-produced books must fight for survival against a flood of material from the U.S. and elsewhere, and even Canadian-produced dinosaur books rarely concentrate on Canadian stories, so have a generic text or surf the dinosaur field worldwide for good stories. The reason advanced by publishers is

Chris McGowan made his own dinosaur out of chicken bones.

that non-fiction requires good illustration, which can be funded only by appealing to the much larger U.S. market.

However, Canadian writers have made some notable contributions to interpreting dinosaurs for young audiences. Reid's and Currie's non-fiction titles (already discussed) were intended for children, and (though filling gaps and perhaps better suited for the adult reader) remind us that many children find answers in books not necessarily written for them. Sylvia Funston's *The Dinosaur Question and Answer Book* (1992) produced in conjunction with the Dinosaur World Tour exhibit, and featuring Canadian and Chinese dinosaurs, has been very successful. Chris McGowan at the Royal Ontario Museum has produced *Discover Dinosaurs* (1992) and (with a new twist on the bird-dinosaur relationships) *Make Your Own Dinosaur Out of Chicken Bones* (1997). And Shelley Tanaka discusses Dinosaur Provincial Park and the *Centrosaurus* bone bed in *Graveyards of the Dinosaurs* (1998).

Simply illustrated materials more strongly focused on Canadian dinosaurs fill gaps that larger publishers have not tackled. Examples include the Tyrrell Museum's *Discover Alberta Dinosaurs* (1992), and Cyndi Smith's locally published *Dinosaurs of the Alberta Badlands*

(1987). Ironically, the Natural History Museum in London feels comfortable producing a full-colour book — William Lindsay's *Corythosaurus* (1993) — on a single Canadian dinosaur.

While non-fiction is usually published professionally, dinosaur-related fiction — with few or no illustrations — is cheaper to produce, and seems to attract the desktop producer, generally with poor results. A few examples stand out for their imaginative or educational content. Using cartoons, Emily Hearn and Mark Thurman's *Mighty Mites in Dinosaurland* (1981) shrinks children so that they are too small to be noticed in a time travel adventure in Dinosaur Park. Perhaps Canada's highest-profile author to write about dinosaurs for children is Mordecai Richler, whose quirky *Jacob Two-Two and the Dinosaur* appeared in 1987. Jacob attends a lecture at (oddly) the Museum of Fine Arts on "The Life of Dinosaurs," whose content is a few decades out of date, and refers only to four dinosaurs from the American Jurassic period. With all of the wealth of Canadian dinosaurs to choose from, Richler has Jacob's parents bring back a *Diplodocus* from Kenya, which improbably has "teeth as large as bananas." Jacob and Dippy escape unlikely pursuers, and Dippy lives happily in the Rockies, stealing pizzas for food.

Geologist Susan Atkinson-Keene's *Weekend in the Jurassic* (1989) seems to be the first fiction to deal with Canada's eastern dinosaurs. Danny finds the jaw of a coelurosaur on his way home from school and

Albertosaurus *skeleton and painted reconstruction.*

(perhaps because he lives in Parrsboro, Nova Scotia) knows what it is. Danny, Kim, their Aunt Ziggy — and their house — travel back to the Jurassic period. There, they encounter a furry pterosaur, a herd of prosauropods, and a small carnivore. A second trip finds hundreds of footprints, which lead to a colonial nesting site and a feather, which shows Danny that feathers came before birds. They return to the present — with a baby coelurosaur.

In the first two volumes of a trilogy, west coast geologist John Wilson combines time travel with another fictional realization of Dale Russell's dinosauroid in *Weet* (1995) and *Weet's Quest* (1997). Aspiring palaeontologist Eric, his sister Rose, and their dog visit the Alberta badlands and find themselves back in the Cretaceous period. The past is brought to life through current research, as the children meet furry pterosaurs and feathered carnivores as well as *Parasaurolophus* and other Alberta dinosaurs. Among other adventures, the children visit a maiasaur nesting ground and (in the second book) travel across the proto–Rocky Mountains to the Pacific coast.

THE EYES OF PALAEONTOLOGISTS

"Artists are the eyes of palaeontologists, and paintings are the window through which nonspecialists can see the dinosaurian world," says Dale Russell. Russell has more experience than many palaeontologists at working with artists (in both two and three dimensions), for he has been able to commission numerous paintings and sculptures to illustrate his books and exhibits.

The first purpose of dinosaur paintings has always been that of representing an animal that is no longer visible in life. The palaeontologist sees a dinosaur from the inside — from the bones — while an artist must present the external appearance of the animal. The bones provide a basis for reconstructing muscles, so the animal's structure can be reconstructed with reasonable success, while rare dinosaur skin impressions allow the texture of the surface of some kinds to be envisaged. Skin flaps and folds, colours and patterns are much more difficult to reconstruct, and a naturalistic posture and setting demands understanding of physiology, behaviour, lifestyle, and environment. As our knowledge grows, it is not surprising that artists' impressions of dinosaurs have evolved.

The pioneer artist of prehistory was American Charles Knight (1874–1953), who was most closely associated with the American Museum for much of his career. Knight made clay models before embarking on his paintings, as many artists have done since. Canadian models

T. rex *is a sculpter's favourite.*

were used for his *Styracosaurus,* and for a Canadian scene prepared for the Field Museum in Chicago in the late 1920s showing *Corythosaurus, Parasaurolophus, Palaeoscincus, Struthiomimus,* and *Edmontosaurus.*

Although Canadian palaeontologists, including Lambe and Charlie Sternberg, made drawings and models, Canada's first major body of dinosaur paintings was created by artist Eleanor (Ely) Kish, who created ten illustrations for Russell's book *A Vanished World,* and fifteen for *An Odyssey In Time.* Born in Newark, New Jersey, of parents who had immigrated from Eastern Europe, Kish knew she wanted to be an artist from her childhood. In the 1930s, she built her own easel in carpentry class, and studied art at night school in addition to regular classes during the day. After years on the road as a painter and craftsperson, she took Canadian citizenship in the 1950s and settled in Ontario. Commissioned to produce a mural for the National Museum, she was asked to draw dinosaurs for a slide show, and met Dale Russell. "Russell realized that in Kish he had met his dream: a talented and disciplined artist capable of translating his knowledge of dinosaurs onto canvas," says interviewer Suzanne Kingsmill. "That was the first time I had chance to paint for a living," said Kish, who quit her job to paint full-time. She makes papier mâché models of skeletons and clothes them with Plasticine muscles before beginning to paint. To get her poses she uses her imagination to put herself in the dinosaur's place. "I'm the dinosaur," she says. "I dream of what I would do, how I would eat, how I would feed my baby and how I would protect myself." She lights the model, makes thumbnail sketches, and then makes a detailed drawing in pencil before beginning to paint.

Russell noted that it took him two weeks to do the specific research required for each painting, while it took the artist about a month to produce a painting. Exploration of questions involved reappraisal of the weight of sauropods, consideration of population density in Dinosaur Park, and devising ways of testing if the atmosphere might have been more dense in the Cretaceous.

As more and more palaeoartists find inspiration in dinosaurs, American painters frequently illustrate Canada's dinosaurs. Baltimore-based Gregory Paul was commissioned by Phil Currie at the Provincial Museum of Alberta to paint his much-reproduced *Centrosaurus apertus herd crossing a stream,* and later Paul created a painting based on an incident recorded in the Peace River footprints. Other artists, whose work has been primarily associated with museum exhibits, are discussed later in this chapter.

HERALDIC DINOSAURS

Early sculptures of dinosaurs in Alberta were discussed in Chapter 6. Dinosaurs seem to offer a particular challenge to the folk artist, who (at any rate, in districts noted for their fossils) is driven to create large — even life-size — models of local and exotic species. For a while, a collection of these models was exhibited in Drumheller under the name Prehistoric Parks, and now the roads through the city are enhanced by such creatures, while the city's rodeo grounds feature a cowboy riding a bucking *Triceratops.*

More serious models are made by sculptors for exhibition in museums, parks, and even zoos. In the 1980s, models were created by Alberta sculptors for the Tyrrell Museum and Calgary Zoo, and government-owned models were borrowed to provide eye-catchers for such promotional ventures as Alberta's Pavilion at the 1986 Vancouver Expo, whose entrance was flanked by a prominent pair of *Styracosaurus* models that have been described as "heraldic dinosaurs." The model of *Stenonychosaurus* (*Troodon*) prepared by sculptor Ron Séguin in conjunction with Dale Russell was discussed in Chapter 9. More recently, visitors to the Tyrrell Museum are greeted at the entrance by a spectacular series of fibreglass dinosaurs.

SCULPTING DINOSAURS

Exhibit plans for the Tyrrell Museum called for a number of life-size models. These were not commodities available off the shelf, nor did Alberta have sculptors with a proven track record in this area. However, Brian Cooley and Mary Ann Wilson, sculptors based in Calgary, had made some small dinosaur models and were asked to produce a life-size *Albertosaurus.* Currie used his footprint research, and "conjured up an image of much more sveldt [sic], sinuous creatures than the plodding, pot bellied behemoths usually depicted."

The sculptors measured a complete, unmounted, skeleton at the Tyrrell Museum, from which they made a 1/10 scale model in polyester resin. Placement of the feet was determined by an actual trackway. Plasticine was used to model the muscles, using information from both muscle attachments on the real bones and living reptiles.

For the life-size model, an armature of steel tubing and rods was constructed based on the real skeleton, and then expanded into a framework on which a surface of burlap and chicken wire could be constructed. A two-inch layer of urethane foam was sprayed on, carved into final shape, and hardened with a layer of fibreglass.

A fragment of preserved skin discovered in the collection at the National Museum at this time allowed the texture of the skin to be modelled. No one, of course, knows what colours dinosaurs were, but Currie felt they would be reasonably well camouflaged. An agreed colour pattern was painted on, then covered by a transparent surface of gelatinous resin.

Polyester teeth and glass eyes were the finishing touches, then the entire skin surface was sandblasted to produce a more leathery appearance.

Using the same technique, other models were made for the exhibit, of the armoured dinosaur *Edmontonia,* and the multi-horned *Styracosaurus.*

Cooley has continued to make models, including one of the new feathered dinosaurs, which was featured on the cover of *National Geographic* in 1998.

DINOSAUR MOVIE STARS

"This is the BBC phoning, from Colorado." It was a quiet Sunday in the 1970s, and I suspected a humorist among my English acquaintances — but no, it really was the BBC. They were making a movie about dinosaurs, and wanted to connect with our team in Dinosaur Park. As scientific interest in dinosaurs grew, so was there more media interest.

Since the Dominion Motion Picture Bureau had filmed George Sternberg in 1921, and the National Film Board had made *Charlie* in later years (see Chapter 5), there was now much greater interest in filming palaeontologists at work. In collaboration with Alberta's educational TV channel, ACCESS, the Provincial Museum made a movie of the *Dinosaur Footprints of the Peace River Canyon,* climaxing with spectacular aerial footage of a series of tracks. Also in the 1970s, the Glenbow Museum produced much raw footage with the assistance of a National Museum grant, some of which was turned into finished films. Attempts to make an IMAX movie of the Canada–China project were abandoned — "it was just too expensive," said Ebbels — but a more conventional movie was made: *In Search of the Dragon.* And the interest from local — and sometimes distant — TV stations increased, leading a steady stream of cameramen to struggle with the dust and shadows of the badlands.

Many more substantial film-for-television series on dinosaurs have now been made, but fiction movies featuring dinosaurs generally are made outside Canada, although they sometimes feature Canadian specimens. The 1925 film version of *The Lost World* included a *Styracosaurus* — it was fighting a Jurassic allosaur, but anything is possible in the movies. Perhaps having more horns than anything else, *Styracosaurus* was a particular favourite; it was filmed for the first version of *King Kong* (ending up on the cutting-room floor) but made it into the 1974 movie based on Edgar Rice Burroughs' novel *The Land that Time Forgot,* and numerous others. A *Chasmosaurus* was the only actual dinosaur featured in *When Dinosaurs Ruled the Earth,* outnumbered by scantily clad cave women. *Ornithomimus* falls prey to a stranded earthman in *Planet of the Dinosaurs.*

In recent years, Canadian resource people are increasingly involved in making raw materials for dinosaur movies. Toronto artist Hall Train has produced dinosaur videos for the American Museum. And in 1993 Michael Crichton's 1990 novel *Jurassic Park* was filmed, a movie that cost, and made, more money than has ever been put into palaeontology. A key set was the group of dinosaur skeletons featured in a climactic scene. Ontario technician and sculptor Peter May built *Alamosaurus* and (from a specimen at the Tyrrell museum) the *Tyrannosaurus* for this group. Other work was done in Canada on the computerized dinosaur animation that makes *Jurassic Park* the most successful popular movie in representing dinosaurs as they are seen by modern scientists.

And now, an IMAX movie has at last been made, with a focus on *T. rex*.

PETER MAY CATCHES UP

Peter May was born near Manchester, England, and came to Canada at the age of eight. "I missed out on the whole dinosaur thing," he says. "They must have taught it after I left England but before I got to Canada." May became a sculptor, and with a degree in fine arts from the University of Guelph, took a job as junior technician in the Royal Ontario Museum's Vertebrate Palaeontology department in 1977, where he learned both lab and field aspects of his work. He moved on to the Tyrrell Museum in 1982, and cast many of the skeletons in the original display, then came back to the ROM as chief technician in 1986. A year later, he set up his own company. As well as movie dinosaurs, May's company, Research Casting International, has made over two hundred dinosaurs for museums and exhibits around the world. The most spectacular is perhaps the rearing *Barosaurus* that now dominates the American Museum's dinosaur exhibits. In 1992 May read a story about Spielberg's upcoming movie and sent a letter about his work.

His casts were not only seen in that movie, but also in an exhibit, "The Dinosaurs of Jurassic Park" that opened in the American Museum the day the movie was released and was seen by 400,000 people within three months. A second set of skeletons and models went on tour in Europe and South America. May's company now has casts in thirty-six museums and other institutions in eight countries, a travelling exhibit touring Canada, and a catalogue of over eighty dinosaurs and many other fossils.

DIORAMAS AND DINAMATION

Modern museum exhibits are the most complex of all artistic creations, as they may include real or simulated landscapes, actual bones and other specimens, casts, models, paintings, sculptures, a variety of audiovisual media, and computer simulations and interactive games. Designing exhibits involves a complex skill known as interpretive planning, and the physical work can involve palaeontology curators, technicians, designers, fabricators, model builders, writers, artists, photographers, and audio-visual and computer programmers.

Earlier museum exhibits in Canadian museums are discussed in Chapter 5, and those of the Tyrrell Museum in Chapter 7. Several other institutions have updated their dinosaur galleries in recent years, but the most striking development has been the opening of new museums or interpretive centres largely related to dinosaur resources. Eastend, Saskatchewan (where Corky Jones lived so many years ago) now has The Eastend Fossil Research Station, featuring "Scotty," Saskatchewan's first *T. rex,* and other local dinosaurs. In Parrsboro, Nova Scotia, the Fundy Geological Museum displays remains of the earliest Canadian dinosaurs. In both, a vigorous field program by staff members is discovering other specimens.

Dinosaurs are also featured in travelling exhibits, which are popular with museums. Dinamation's exhibit "Dinosaurs," came from the U.S. but featured dinosaurs found in Canada — *Triceratops, Pachycephalosaurus, Parasaurolophus,* and *Tyrannosaurus.* The half- and full-scale robotic models "lift their heads, swish their tails, stamp their forelimbs and roll their eyes" while roaring and grunting. Dinamation robots have a permanent place at Sudbury's Science North. They were popular at the ROM in 1988, when a planetarium show, "Death of the Dinosaurs" (originally developed by the Vancouver Planetarium), was featured. The Provincial Museum of Alberta (with the aid of several sponsors) displayed another Dinamation exhibit, "Carnosaurs," in 1995, which brought in sixteen-thousand visitors in its first two weeks (of a projected 175,000 visitors for a six-month run).

The imagination displayed in *Jurassic Park* and its sequel, *The Lost World,* together with the success of the Dinamation exhibits has led to some criticism from professional palaeontologists, but the public seems happy. "Carnosaurs has a strong educational focus, but there's no denying this is wonderful entertainment too!" says Provincial Museum staff member Kathleen Thurber.

VLADIMIR KRB — PALAEOARTIST

"Palaeoartists are a unique breed. They have the curious, inquiring mind of the scientist, and the observative, sensitive, creative nature of the artist." Vladimir Krb was born in Czechoslovakia, and became interested in both art and science as a child. He remembers visiting the National Museum at the age of ten and being enthralled by the fossil trilobites there. At fifteen, he visited Zdenek Burian, a palaeoartist of international renown whose illustrations (including Alberta dinosaurs) have turned on many young fossil enthusiasts. At first Burian's student, he became a collaborator in later years. But things were difficult in Czechoslovakia, and eventually Krb went on holiday to Austria. "He took his wife, his two-year-old son, and two suitcases. He never returned. All his rights to his work were forsaken." He came to Canada, and in 1983 became the principal artist at what would become the Tyrrell Museum.

To create a mural Krb prefers to work first with the scientists, obtaining the hard facts on which his reconstructions must be based. He then gathers information on the dinosaurs' environment. The next step is a "clay model, with all the elements of the skeleton and musculature, and the proportions and stance taken into consideration." From this model, accuracy, position, and the play of light can be evaluated. The galleries were modelled on a small scale, and Krb lay on the floor for hours playing with lighting and models. "A mural requires collaboration between only one good scientist and one good artist," says Krb. "He provides scientific information, I provide the artwork." The paintings were done in oils, photocopied on transparencies in Ottawa, and then digitized in London, England. Then, computer-controlled airbrush spray guns applied the colours onto a cotton canvas rolled on drums — a huge "paint-by-numbers" method that produced the final murals.

Krb's paintings are an important feature of the gallery, adding information and atmosphere to the specimens, showing them in life when the "real thing" is a skeleton. In many examples, his paintings place the "living animals" in the same positions as the skeletons mounted in front of them. Particularly important is the diorama's ability to counteract misunderstandings, for the dinosaurs are seen in a natural habitat instead of in the badlands scene so many visitors expect. One features a "graveyard of drowned centrosaurs," while another provides an underwater seascape surrounding the marine reptiles.

BEAVER, MAPLE LEAF . . . AND ALBERTOSAURUS?

In terms of public interest and excitement, dinosaur study is one of the most important areas of science. Canada has some of the richest dinosaur resources in the world, and — at least in the last three decades — it seems that Canadian science and interpretive institutions have begun to do them justice. Dinosaurs are beginning to be featured with a degree of excellence in the most realistic and directly interpretive art forms of non-fiction writing, representational painting and sculpture, and museum exhibition. Yet, although Canada's creative artists in other media have certainly taken an interest in dinosaurs, it is striking that — outside fiction for young audiences — much of their work has not been inspired by Canadian dinosaurs, researchers, or discoveries.

T. rex *guards Drumheller arena.*

Canadians who have been inspired by Canadian dinosaurs and dinosaur discovery to move beyond the facts tend to be those with their feet firmly on the ground in or near the dinosaur localities — often (though not always) in the west. Robert Kroetsch, Al Purdy, Monty Reid, and John Wilson are western residents; Susan Atkinson-Keene is in the east. The "big names" of our culture, centrally based, seem to operate with less understanding and awareness of the real basis for their dinosaur-related stories.

In fact, it seems that many Canadians — if they think of dinosaurs at all — see them as only a regional reality, a hinterland product, like polar bears, cod, and salmon — other neglected natural resources from the periphery of the country. Perhaps if dinosaurs occurred naturally in Ontario and Quebec, Canadians in general would take them more seriously, and we might even see *Albertosaurus* or *Edmontosaurus* join the beaver and maple leaf as a familiar Canadian symbol.

13

SUPPORT YOUR LOCAL DINOSAURS

"I've gone far enough into the field now to realize that, no matter how long I spend in it there's always going to be more mysteries opening up in front of me, and I'll never know all there is to know about dinosaurs."

PHIL CURRIE

DINOLAND IN DRUMHELLER

North American popular interest in dinosaurs is so high that Dinoland opened as part of Disney's Animal Kingdom Park in Florida in April 1998. "When looking for ways to immerse staff in the world of dinosaur palaeontology, Disney looked north to the Tyrrell Museum," said the Tyrrell's newsletter *Trackways.* "For five days, 10 performers learned to cast and prepare fossils, examined the spectacular fossils Museum staff are preparing and toured the museum from top to bottom. . . . On . . . a Dinosaur Provincial Park hoodoo . . . Director Mark Renfrow . . . pulled out his cell phone and called his boss in Florida to rave about the experience."

The ten performers who trained at Tyrrell will "play palaeontologists exploring the world of dinosaurs." Their presence at the Tyrrell Museum is a compliment to that museum, and to the quality of its staff and programs — but, ironically, the number of performers being trained to "play palaeontologists" is around five times the number of professional dinosaur palaeontologists working in the whole of Canada.

Yet the number of people in Canada who owe their jobs to dinosaurs (directly or indirectly) has never been higher. While the Tyrrell Museum and Dinosaur Provincial Park undertake many tasks not directly connected with dinosaurs, there can be little doubt that these institutions would not exist without Alberta's high-profile dinosaur resource. The Eastend Fossil Research Centre and Parrsboro's Fundy Geological Museum also both reflect the dinosaur presence. Although only part of their overall activities, dinosaurs are studied at the Canadian Museum for Nature and the Royal Ontario Museum, and to a lesser extent at some other museums. Several Canadian universities now support academic staff working at least in part on dinosaurs, and they in turn teach graduate students, some of whom are preparing theses on dinosaurian topics. And in the last few years, a number of businesses have become established that depend substantially on dinosaurs. Since Lambe began to focus on dinosaurs early in the century, the level of directly related employment they generate has increased from one to perhaps a hundred people. And there are many more at least partially supported in spin-off businesses such as accommodation, food services, transportation, education, and publishing.

This increase reflects a remarkable change in the economic importance of dinosaurs in Canada. Since we are most willing to pay for what we value, the flow of dollars reflects the most direct connection between dinosaurs and our society — yet strangely it is a topic that is not often discussed in books on dinosaurs, which like to pretend that science is independent of economics, and that marvellous discoveries are not in any way dependent on dollars.

DINOSAUR RUSTLERS AND EXCESSIVE REGULATION

Since people are interested in dinosaurs, dinosaur bones are worth money. "The rock-hounds are going out with half ton trucks and removing great quantities of bone," said Potter Chamney at the 1963 conference. "I would like to . . . find suggestions for legislation on the removal of vertebrate remains from the Badlands." Many rockhounds are hobbyists, some are collectors who want to work legitimately with dinosaurs (but feel that dinosaur bones are just as much a resource for business as gravel or coal), and there is a thriving black market in dinosaur and other fossil material, serving primarily the needs of millionaire collectors, for whom a vertebrate fossil is a spectacular conversation piece.

In Canada, federal legislation protects resources in national parks and prohibits export of certain specimens, but otherwise the responsibility is in provincial hands. Alberta's Environment Conservation Authority held hearings in 1972 into protection of archaeological and historic resources in Alberta, and strongly recommended protective legislation. Palaeontology was not included in the terms of reference, but some speakers did urge fossil protection.

"The worst fears of university researchers in the 1970s were realized," said Mark Wilson at the University of Alberta, when, without consulting the researchers who were to be affected, the government introduced a system of permit requirements for palaeontological research. The 1973 *Heritage Resources Act* (which has become a model for such legislation) declared that fossils were the property of the province, allowed surface collecting in areas outside parks, and instituted a system of permits for legal excavation by palaeontologists and fines for ignoring the provisions of the act.

Fossils were added to the bill at the drafting stage, without consultation with the government staff working in palaeontology, so that Provincial Museum staff were placed in the embarrassing situation of being expected to implement an act that understandably annoyed their university colleagues. An advisory committee was set up, and subsequent revisions to the act smoothed some of the most irksome provisions, but some objected to the principle of provincial control, and have argued cogently that the act presented "a case of excessive regulation."

Bruce Naylor, now director of the Tyrrell Museum, was then a graduate student at the University of Alberta, and "saw it as a government intrusion that might limit my access to fossils that I needed for my research. . . . Today I find myself helping to administer the very law that worried me so much. . . . I believe the overall impact of the law has been very good. Scientists and museums have been able to continue to collect freely, under permits reviewed by a panel of our peers." However, some outside scientists are discouraged by the requirement that material should return to the province after study. "Some people don't even try and work in Alberta because of the legislation," says Currie.

No special funds were provided for enforcement, and collectors without scruples or knowledge continued to pillage the fossil beds as they always had done. Once removed from its site, it is hard to prove that a dinosaur bone was excavated rather than collected on the surface, but there have been successful prosecutions. In 1987 a Drumheller rock shop proprietor was fined $2500 for stealing dinosaur fossils from Dinosaur Park. The value of fossils is leading illegal commercial collectors into occasional violent confrontations in the badlands, and in 1997 teeth were twice stolen from an *Albertosaurus* skull already plastered for the Tyrrell Museum.

Fuelled by Hollywood, Dinosaur Provincial Park attendance has nearly doubled in the last five years. The growing interest in dinosaurs gives park staff the opportunity to spread the message more widely, but it also brings in more undercover collectors. "One more dinosaur movie," said ranger Brian Bennet, "and this whole place could blow wide open."

Educational needs of the legislation were served indirectly, as the Provincial Museum and then the Tyrrell Museum raised the profile of dinosaur research based on new discoveries, and more directly as media and speakers to rockhounds and other groups presented the message that dinosaur (and other) fossils were important, and that a bone turned into a tie pin or sitting on someone's mantelpiece can deprive science of the opportunity to fill some vital gap in knowledge, perhaps for ever.

DINOSAUR DOLLARS

In 1993 the Dinosaur World Tour gift shop sold "dinosaur dollars." With a *Mamenchisaurus* on one side and the dinosaur logo on the other, the simulated coin is a favourite souvenir for many people. Dinosaurs have financial value beyond their black market price. We contribute real dinosaur dollars through taxes, donations, admissions, and corporate sponsorships to those institutions and individuals studying and interpreting Canadian dinosaurs. Though no comprehensive study is available, and some rather arbitrary decisions will be needed to distinguish dinosaur projects, I have attempted to make a ballpark estimate of the amount Canada spends on its dinosaur activities, and to examine how museums are surviving lean times.

The biggest institution in Canada that is strongly dinosaur-related is the Royal Tyrrell Museum, which is an Alberta government department. Its capital budget was around $30 million, and its annual budget currently runs to nearly $3 million. Like all government enterprises these days, the Tyrrell has suffered cutbacks; the budget declined substantially in 1991 and bottomed out around 1995. The government contributes about half of the budget for salaries, and Public Works maintains the building. "But program delivery and exhibitry depend on admission money and corporate sponsorship," explains Director Bruce Naylor.

Field trip in Dinosaur Park.

The Tyrrell Museum does not raise chickens and repair cars like the Chinese IVPP, so voluntary support has become important. The museum's Friends managed donations until the government introduced an admission fee. The present cooperating society manages admissions and services. Figures for 1997 show the top five revenue earners as sales ($1.4 million), admissions ($780,000), programming ($305,000), publications ($96,000), and cafeteria operation ($69,000). However, some of the revenue goes to cover the costs of raising money; the gift shop, for instance, has to pay staff and the cost of products, so revenue is really nearer to $500,000 ($513,000 in 1997).

Non-government funds are essential to the Tyrrell's operations. "We raise about one million a year in admissions, and another half million in donations and corporate sponsorships," says Naylor. The public also contributes directly to popular programs. "Through the public's 'pay to dig' program, our Dinosaur Park Fieldwork program has become self supporting."

Of the Tyrrell's expenses, a substantial amount — some $380,000 — goes to dinosaur fieldwork and preparation. Funds for major projects and overseas travel have to be raised separately. "Dinosaurs can raise money, so my program gets cut," says Currie. "Then the government cuts the museum for the same reason. Somehow we've survived."

Canada's other major program has been that at the Canadian Museum for Nature in Ottawa. "The national museum is a very sad situation," commented a museum official at another museum. "Through policy decisions one of the great museums has now just got interpretation without curatorial support, and now has not got interpretive resources." Dale Russell is still listed as a Curator Emeritus, but he left the Canadian Museum of Nature several years ago and went to North Carolina, where he continues to write joint papers with Currie on small carnivores. "Budget cuts decimated the curatorial programs," commented another colleague. "Russell had to apply for technical time and then find the money for it."

Although I had had enjoyable meetings with curators on previous visits to Ottawa, in a 1997 visit I was unable to connect with any curatorial staff for an interview or get any information on research programs. "It's not necessarily irreversible," I was told elsewhere, "but they have to bring in people with competence and willingness to do something new. They have excellent collections, but no research. If they just have an exhibit, they forfeit their claim to be Canada's national museum of natural history." Michael Caldwell joined the museum in 1998 as a research scientist with a McGill Ph.D. and wide interests in

Hans-Dieter Sues.

fossil reptiles, while research associates Robert Holmes and Cathy Foster are describing a new ceratopsian dinosaur.

Now, the principal central Canadian museum working on dinosaurs is the Royal Ontario Museum. With its many departments, palaeontology represents only a small part of the museum's budget. Hans-Dieter Sues is Senior Curator of the Department of Palaeobiology. "We've done relatively well despite the cuts," he comments. "We do important work, we're visible. Museums are very polarized — the people on the public side feel they are the only relevant thing. But the public comes to the museum to see dinosaurs, so it's important to have curators, to provide the information and generate additional material. The old-style curators saw the public as a nuisance, whereas many of the new curators are there because they enjoy being educators."

A much smaller facility, the Fundy Geological Museum at Parrsboro, Nova Scotia, has an annual budget of around $300,000, to which grants and special exhibit funds are sometimes added.

It seems then that Canada provides public funding of somewhere under $1 million — the price of four average Toronto houses — directly on dinosaur fieldwork, research, and preparation per year. Significant contributions to the total are made by the federal government (national museums; museum, economic development and academic grants), provincial governments (provincial facilities), corporations (project grants), and directly by the public at large (admissions, fees, and purchases). If all public activities in some way related to dinosaurs are included (exhibit buildings, interpretation programs, university courses, etc.), the total might be $5 million. A small amount also comes from outside the country — funding for visiting researchers, and occasional grants from the National Geographic Society and the Dinosaur Society to researchers working in Canada.

THERE'S GOLD IN THEM THERE DINOSAURS

"The Tyrrell Museum of Palaeontology . . . has been far more successful than the forecast made in the planning stages," says a study conducted by Walter Jamieson of the University of Calgary, published only three years after the new museum opened its doors. Drumheller had been suffering from declines in oil, gas, and agriculture, and without the Tyrrell Museum economic activity would have been down thirty percent. Jamieson's interviews documented the creation of thirty-two new businesses, and showed that "the facility . . . has given Drumheller a new image and civic pride."

When in 1947 the Geological Survey of Canada transferred its dinosaur program to the National Museum, it was clear that dinosaurs were not regarded as in any way related to the Survey's priority, the pursuit of minerals of economic significance. Yet the resources of coal, oil, and gas that the survey had been instrumental in discovering had ceased to serve Drumheller well, and the city needed a new economic generator. Some of the new revenue to the community came from a shift of government expenditure — dollars paid in Drumheller that previously would have been paid in Edmonton, or not at all. But the bulk of the profit came from new dollars attracted from outside by the museum.

The possibility of using dinosaurs to generate revenue came through the evolution of new directions in tourism — cultural tourism (which uses history and culture as its generator) and ecotourism (the drive of the traveller to seek out authentic, life-enhancing experiences, often in remote places). Dinosaurs fit in both areas, as a resource that could be exhibited in a major museum, and yet one that comes from wild landscapes.

Historically Alberta's resources had been ranching and farming, coal, oil, and gas. Yet by 1992 tourism generated $2.95 billion in revenues, making it Alberta's third-largest industry. Over half of this was generated by travelling Albertans, another $726 million by other Canadians, and nearly as much by visitors from other countries. Part of this revenue reflected a government decision to spend in order to recoup — to build attractions that would attract tourists, who would in turn leave dollars behind in the communities they visited.

In Alberta, part of this spending has been in the development of a network of major historic sites and museums. Around $100 million petrodollars from Alberta's Heritage Fund have been invested over a period of twenty-five years in such facilities as Head-Smashed-in Buffalo Jump, the Cochrane Ranche, the Reynolds-Alberta Museum (featuring historic transportation and agriculture), and the Tyrrell Museum. Studies have shown that each of these facilities paid for itself as an economic generator within a few years. The Tyrrell attracts an attendance equivalent to about half of the total for all of the province's historic facilities. "There has been $7–9 million in direct economic spinoff since the Tyrrell was built," Director Bruce Naylor told me in 1997.

This success has led other provinces to look at their dinosaurs as economic resources. The Royal Saskatchewan Museum's discovery of a *T. rex* in 1994 led to the development of the Eastend Fossil Research Station, a joint venture between the museum, the Government of Saskatchewan, and the Eastend Tourism Authority. There "Scotty" is being prepared under the eyes of some six thousand visitors a year, and other fossils from the area are on display.

At the other end of the country, in Parrsboro, Nova Scotia, the Fundy Geological Museum has been set up to display dinosaur and other geological material from the Triassic and Jurassic periods, and attracts over twenty thousand visitors a year. A number of other museums (though not undertaking dinosaur research programs) find it advantageous to display dinosaur casts or models.

PUBLICITY WORTH $50 MILLION

Great international travelling exhibits have been more cash-intensive, both in initial funding and in anticipated revenues. The 1990 Hitachi-Dinoventure in Japan cost $1 million, while Alberta Tourism hoped that "it may generate publicity worth $50 million in advertising dollars, and bring Japanese tourists to the Badlands of Alberta."

The great experiment of the Ex Terra project cost a lot more. Initially, many of the research dollars were spent in China, not Canada, and were investments in discovery and science, while the expected economic gains were to be reaped through the exhibit. Early announcements referred to the three-year travels of the King Tutankhamen exhibit, which had earned $16 million for its Egyptian sponsors; with expressed interest from twenty museums around the world, it was clear that expectations for the Dinosaur World Tour were higher. When the exhibit was announced, the accompanying pamphlet outlined potential benefits to the province and country through direct employment, increased international trade in Canadian exhibit and communications technology, direct exhibit fees, sale of associated publications and media rights, and the promotion of dinosaur fossil display centres and collecting regions as international tourist destinations. By 1989 Ex Terra expected to have a cash-flow shortfall of $9.3 million, and asked for loans from the federal and provincial governments of $6 million each. Projected direct economic benefits from the exhibit were "over 875 man-years of employment and more than $40 million in export revenues and tourism revenues for Canada."

When I spoke to Ex Terra Executive Director Jim Ebbels in 1997, he told me that the project had not had any government money since 1993. "We've been living off the proceeds of the exhibit and merchandising since then — and we're still here. There are many detractors, but the governments are pretty happy, and the exhibit has been quite a big hit in the local economy. [In Edmonton] we spent three quarters of a million to generate fifteen." Ebbels was still optimistic that the exhibit would continue, and that Ex Terra would take on new projects of the same kind, but later in the same year the Ex Terra Foundation closed down, and its government-owned assets went to the Tyrrell Museum (which — after initial suspicion — had provided so much support).

Although the exhibit was seen by more than 2.5 million people, it certainly did not make the multiyear tour in four countries that was envisaged. "It met its objectives, and promoted the science," says Tyrrell Director Bruce Naylor. "The museum benefited from the research and fieldwork in China. And it hit the market — Japan was a great success."

Although no formal evaluation has been published, it seems likely that the large scale of the Ex Terra exhibit made it impossible to mount in the major museums that were at first interested. "We couldn't accommodate it for lack of floor space," said Sues at the ROM. The huge amount of volunteer support required to run it successfully also must have been a major barrier to some communities.

The tantalizing thought exists that a less ambitious and more concentrated exhibit, which still would have featured the new finds and exciting approaches, might have been attractive to the many museums in the U.S. and other countries featuring dinosaurs, and could have made its multiyear target. Certainly something of the kind is in the minds of the Tyrrell staff, who in 1998 were busy planning a smaller exhibit in conjunction with the Indianapolis Children's Museum. "We hope it will tour the States and then go into Europe," said Naylor.

To some, the Dinosaur World Tour — and perhaps the whole Ex Terra project — was a failure, and perhaps it *was,* in terms of its own ambitions. Yet the research component was an undoubted success, doing much important work and linking Canadian scientists across the Pacific Ocean in a way that has continued to bear fruit. It also produced probably the most spectacular dinosaur travelling exhibit the world has yet seen, and showed off a major Canadian international scientific initiative to a substantial audience in four countries. Its more abstract successes are hard to quantify: Can we put a dollar value on Phil Currie's face on the cover of *Time,* or tell what will bring tourists to Dinosaur Park in ten years' time? Perhaps some critics of Ex Terra belong to a national tradition that says if a project is too imaginative and ambitious, Canadians shouldn't be doing it. Could this be the Avro Arrow revisited?

One spinoff of Ex Terra is the number of dinosaur-based businesses that have developed. As Canadian expertise has built up in the production of new museum exhibits, spinoff businesses have developed that serve museums in other countries. Former Tyrrell technicians Gilles Danis and Peter May are both running dinosaur casting companies, while in 1990 Ely Kish completed a 5 x 43-metre (16 x 140-foot) mural for the Smithsonian.

Canada is selling expertise in other ways, too. Knowledge about dinosaurs, exhibit design and interpretive planning, model making, and dinosaur tourism are all saleable commodities, and in various ways Canadians have been marketing the expertise generated through dinosaur projects in the country and outside it. A number of the talented group of people that gathered around the Provincial Museum and the Tyrrell, Ex Terra, and the university programs of vertebrate palaeontology are now working in a variety of fields and places — as researchers in the U.S. and Canada, as exhibit builders in Canada and the U.S., and hosts of educational TV shows.

"We cannot forget that the primary objective of the process we are discussing is the preservation and interpretation of cultural resources," concludes Jamieson in his economic impact study of the Tyrrell. And indeed, although it is important to recognize that dinosaurs are of economic significance, we must never forget that their importance in education and science is, ultimately, of more importance in our society than the dollars they bring in.

AN OUNCE OF CURIOSITY

"Of all sciences, paleontology is the easiest to open the door for anyone with an ounce of curiosity," said Ex Terra's Jim Ebbels. Dinosaurs have a particular appeal for the young, and they also offer great educational advantages, allowing children to enter such important topics as astronomy, behaviour, ecology, evolution, and extinction without being discouraged by too many complex concepts. How does a Canadian child find out about dinosaurs in his own country? Teachers often use dinosaurs in today's classrooms, working them in anywhere there is room — in language arts as well as in science. Yet the resources we offer to teachers and children are not always the best.

The finest experience is to work on a dig, where dinosaurs are seen in their natural context. The Tyrrell Museum organizes a variety of programs in which digs can be experienced or observed. Other contact with the real thing comes through museum visits, and there is now somewhere in almost every province where a child can see something dinosaurian — at least a skeleton or a model. British Columbia has a skeleton in the geology museum at UBC, though the rest of the province has less to offer — a single tooth in the Courtenay museum and a few tracks at the Royal B.C. Museum and in Hudson's Hope. In Alberta, Edmonton, Calgary, and Drumheller each provide experiences. Saskatchewan has skeletons in the university in Saskatoon, "Megamunch" in Regina, and "Scotty" at Eastend. There are dinosaurs on display in Winnipeg, and many in Toronto and Ottawa. In the Maritimes, Parrsboro and Halifax can both show material. Quebec, the other Maritime provinces, and Newfoundland seem to be without public dinosaur exhibits. In the face of reduced school budgets, however, many schools are unable to visit dinosaur sites, and children depend on parents for field trips.

The next-best resource should be the school library. A number of Internet resources are now available on dinosaurs from Canada and elsewhere, and naturally one would expect school libraries to have the best of books on this popular topic. During school visits, I check the library, and sadly have to report that many are disappointing. Cutbacks are affecting library budgets, and even more serious is the disappearance of

A hands-on dinosaur dig at Dinosaur Park.

teacher-librarians (who have the expertise to choose good materials) from many systems. Books on the shelf tend to be out of date, and frequently do not include any Canadian content at all, as American books are generally cheaper and better illustrated.

A related problem is the mere availability of books for children on Canadian dinosaurs. Some excellent titles have been published, but few feature Canadian material extensively. Several Canadian publishers I have talked to assure me that there would not be a big enough market for a children's book on Canadian dinosaurs, though bookstores, parents, and teachers tell me they would love to have one available.

Much of what kids learn about dinosaurs is from television, which features them in science shows and general-interest programming. Yet again, most of the material comes from elsewhere — usually the U.S. — and naturally features American scientists rather than Canadian ones. That was the situation when Phil Currie was a child, and the only role model he could find in dinosaur collecting was American Roy Chapman Andrews. Yet, ironically, Currie now has his own following.

"Phil [Currie] is a pop star in Japan," said Jim Ebbels. "When the [Dinosaur World Tour] was in Osaka there were 20 fax machines in the exhibit. They did an educational TV show, and kids from all over Japan could fax their questions to where Phil was walking through the exhibit with the host." If Currie is a "pop star" to Japanese children, what is he to Canadian children? Despite frequent involvement in the media, Currie is too far away from centres of production to appear in many Canadian programs for children. Yet he is now a big enough name that he is sought by the Discovery Channel and the Learning Channel — "I'm on camera way too much," says Currie. Even if his image is carried through U.S. media, it looks as if Currie will continue to be a role model for Canadian kids.

MORE MYSTERIES

While dinosaurs are important for economic and educational reasons, the real justification for dinosaur study is the scientific pursuit of disinterested curiosity. Dinosaurs are studied for the same reason that people climb Mount Everest — because they are there. After a century of dinosaur research in Canada, we have not reached a plateau of knowledge — rather, we are breaking through a barrier to new knowledge, and everything ahead of us seems rich and strange. Much that we thought we knew is not entirely true, much that we didn't know is becoming clear, and new knowledge brings new questions. The pace of recent discovery shows clearly that there is much more to be found, and I have not talked to any researchers who are short of ideas or projects.

At the Fundy Geological Museum, Tim Fedak is preparing the prosauropod collected in the summer of 1992. Already exposed is a humerus and hand, the pelvis, both feet, and part of one leg. The animal is lying on its side, and there is no sign of a skull. But Fedak has also been spending time along the cliffs. "He found a second animal 4 metres away and went looking for more bits," says Director Ken Adams, who with other staff members has now put in one hundred hours of searching for further material. Despite the difficulty in finding bone, "there are five or six animals recognized from the site." The history of western discovery is now being repeated in the east, where locally based staff making regular site visits are able to find bones that have been missed by visiting researchers for decades. "This is the real material, all jumbled up," says Adams, "but the kids expect big skeletons. *Jurassic Park* gave them eating, breathing, stomping dinosaurs — this is the real Jurassic Park."

In Alberta, Tyrrell's Currie is continuing to write joint papers on the small carnivores with Dale Russell. He is excavating an unusual bone bed containing carnivore remains, an *Albertosaurus* site first found by Barnum Brown and then lost for years. "I saw the material in the American Museum — nine right feet, all from one site . . . a tail and some lower jaws. And I found four photos labelled 'Albertosaurus quarry'. [A team from the] Dinamation [company] went down the Red Deer. They tried to recreate Brown's tracks and found the quarry. It's only about a quarter excavated. It's the first evidence of packing behaviour in tyrannosaurs." The young Albertosaurs were very lightly built — "Brown thought they were Ornithomimids."

Currie's colleagues, Don Brinkman and David Eberth, are continuing other bone bed studies, while palaeobotanist Dennis Braman is continuing his studies of the K-T boundary, at sites in Alberta and down as far as Mexico, seeking further understanding of the nature of the changes that took place at this much-discussed turning point in Earth's history. Technician Darren Tanke has carved out his own research niche: palaeopathology, the study of dinosaur diseases. Handling many bones, he was struck by the number that showed signs of ancient injury or disease, and began to gather data, which now appear on his own Web page and in various publications, some with collaborators from modern medicine. Perhaps his most fascinating discovery was a *T. rex* with gout — a painful disease traditionally associated in humans with (among other things) consumption of red meat. Tanke does not speculate on the disease's possible impact on *T. rex*'s temper.

In Saskatchewan, site of Canada's first scientific dinosaur discoveries, research had been in the hands of local amateurs like Corky Jones and visiting specialists until the 1980s. Then the Royal Saskatchewan Museum created a lab, put together by Tim Tokaryk, who — while still at school — started out as a volunteer with the Provincial Museum of Alberta and then worked as a technician at the Tyrrell. In 1990, he took a new position of assistant curator of fossil reptiles and birds with curator John Storer. He has found a *Triceratops* skeleton, as well as numerous vertebrates that are not dinosaurs. In 1991 he participated in the discovery of the province's first *T. rex*, and in 1995 (with Wendy Sloboda, discoverer of the Devil's Coulee egg site in Alberta) found the world's first *T. rex* coprolite, thus receiving the dubious honour, he says, of "having worked on both ends of a *Tyrannosaurus.*" Tokaryk now runs the Eastend Fossil Research Centre, preparing the *T. rex* by day, and in his spare time studies the small theropods that have been found close to the K-T boundary.

German-born and trained in Alberta and the U.S., Hans-Dieter Sues at the ROM started out his career with global interests. He became interested in the dinosaurs of Nova Scotia while he was at Harvard. "The Triassic exposures are not very rich," he says with understatement. "I get about three fragments of bone per mile of coastline." Later material offers promise: "The early Jurassic dinosaur tracks show quite a few small theropods," he comments. "I hope one of these days we'll start pulling theropod dinosaurs from the cliffs."

Mining the collections in the ROM's basement is more productive. "I found the edentulous maxilla [toothless upper jaw] of a theropod," says Sues of a plastered Alberta specimen collected by Lindblad in 1923 and never opened. "I was so excited, I opened the other block. There was a pelvic girdle, additional vertebrae — it was a new type of theropod." The technician preparing it thought he had an extra vertebra beside the pelvis, "but I realized I had a braincase," says Sues. "I was so delighted, I almost jumped up and down." These fragmentary specimens have teased palaeontologists for decades. "Small theropods are the most interesting thing in the Cretaceous," says Sues, "yet they are represented by the scrappiest evidence." Naturally, Currie, too, is excited by the find: "It's the best specimen yet," he says.

NEW FRONTIERS

Tyrrell's connections with China continue — "and now they can hardly keep pace with the general reporting of discoveries flooding out of China," reports Louis Psihoyos. There will be ongoing studies with particular reference to feathered dinosaurs and the origin of birds, but Currie is uncomfortable with the controversy. "There's more and more speculation published," he says. "I like to present the facts and deal with them, and that's that." Despite the frustrations, Currie is still excited by his work. "I'm still having a lot of fun with it," he says.

Meanwhile, in 1997 Currie, his wife Eva Koppelhus, and associates Don Brinkman and Mike Getty began working in Argentina with Rodolfo Coria. In the Patagonian desert they are excavating several large theropods related to *Giganotosaurus*, a carnivore that may be several tonnes larger than *T. rex.* (Coria is also working with Currie in the Alberta theropod bone bed.)

Sues is interested in the way dinosaur and other reptile faunas responded to the breakup of the supercontinent Pangaea, which has led him to study a number of sites abroad, including Morocco and Yunnan, China. He is currently interested in the Kyzylkum Desert in Uzbekistan, a breakaway state from the old U.S.S.R. "Unlike the Gobi desert," Sues explains, "it represents a lowland floodplain near marine settings as in Alberta. From the fragmentary evidence of Russian researchers it resembles Western Canada, but it's older."

What other discoveries are possible? A Canadian sauropod would always be nice. And Currie has long wondered about the possibility of a pickled dinosaur in Alberta's tar sands. Research will continue on the question of hot-blooded dinosaurs, and other researchers will continue the "Jurassic Park" search for the elusive dinosaur DNA. While some directions of research can be predicted, there are always surprises. Every museum basement contains specimens that could not be understood when they were found, or that remain unprepared through shortage of staff and funds, but that can now be interpreted. Other new and exciting specimens that will change our understanding are waiting below the surface in the badlands, and every rainstorm brings them a little closer to discovery. And in Canada's eastern dinosaur area, Donald Baird reminds us: "Thanks to the inexorable Fundy erosion, collecting in Nova Scotia can go on forever, with worthwhile stuff turning up every year if there's somebody on hand to chisel it out."

AN INTELLECTUAL JOURNEY

Canadians have taken many journeys through their land. We have sought wealth in skins, crops, gold, and oil; we have sought to shrink its vast size by building railways and roads and by travelling beyond their limits in boats and planes. Most fascinating are the intellectual journeys into the far corners of the land to collect shells and rocks, to learn new languages, and to find old stories. To some Canadians, journeys in search of past life have been the most interesting. The country is rich in evidence of early human life, and even richer in evidence of prehuman life forms. Among many sites, we have some of the oldest signs of life of any kind in Ontario's Gunflint cherts; the richest Cambrian site anywhere in the world in the Burgess Shale; and the oldest representatives of several vertebrate groups. Is it fair that we also have the world's fourth-richest dinosaur fauna, and its best single dinosaur site?

Dinosaurs are by no means the only fossils of scientific importance, but they have come to have a magic that few other fossils — apart from the very rare remains of our own kind — possess. They seem to symbolize the excitement — and the remarkable results — of the quest for past life. In telling the story of Canada's dinosaurs, we have seen how this rich resource was discovered by First Nations, then by European and Canadian scientific explorers. After a period in which we lost control of our resources and scientific leadership to the U.S., Canada has taken steps to protect its resources, and has hired and trained palaeontologists to study them, in partnership with collaborators from other countries. We no longer export dinosaurs — now we can export expertise.

Within Canada, we have seen a shift from fossil science dominated entirely by national and central institutions, to one significantly supported by provincial initiatives in areas where dinosaurs are important to local education and the economy. We have built important institutions to study and interpret our dinosaurs and other fossils. And with Ex Terra we stumbled — in a very ad hoc Canadian fashion — into a major international effort through a combination of individual and institutional creativity, supported by a mix of federal, provincial, corporate and cultural funding, and volunteer support. We have brought at least some of it off triumphantly, and given Canadian dinosaur science an international profile that we do not have in many other fields.

As this century rushes to a close, Canada is going through a major social and economic upheaval. Long-cherished Canadian values — medicare, education, culture, heritage — are under siege. The study and interpretation of dinosaurs is one of many areas that suffers as public funding is cut back for major institutions and programs. Some have been so threatened that their basic functions have been put in jeopardy. "People don't often appreciate that museum collections are not ours," says curator Chris McGowan at the ROM. "We are looking after them, they'll be available for ever we hope, to consult. When there's a new

Jurassic Park everyone wants to know. Where would people go for their information, for the significance of recent finds? I see the point of having an entertainment-type approach — but to preserve collections is of vital importance."

"Support your local dinosaurs," says the Tyrrell Museum guide. There is no objective standard that determines how much effort should be put into a scientific program, but surely Canada can afford to support more dinosaur palaeontologists than Mongolia or Argentina. "We could use another dinosaur palaeontologist," says Bruce Naylor at the Tyrrell, with serious understatement. And as one of the richer countries in the world, we should be able not only to study our own dinosaurs, but also to continue to be meaningful contributors to international science on the world stage. There is no shortage of opportunity.

Dinosaurs have begun to play a part in Canadian culture, but here, as so often, we are still dominated by influences from our large neighbour. We are pleased when we get back into Asia ahead of the Americans but are flattered when Disney comes to the Tyrrell Museum to learn. Our dinosaurs should appear in our popular culture as well as our poetry; the profits from a Canadian movie on the scale of *Jurassic Park* could pay our entire dinosaur science bill for the next century — if Canadians could afford to make such a movie and reach the markets to sell it.

Economic pressures are always pushing dinosaur interpretation into showbiz. While there are those for whom Disney-style presentation is better than reality, many people can tell the difference. "What the public wants to do is to go to a place where they can see authentic objects that reflect on their own history and place in the universe," says Hans-Dieter Sues. "To handle a dinosaur bone is a magic experience — almost like a religious experience."

For Canadians, dinosaurs have come to be poetry and pizazz, recreation and revenue, education and science. Although the bones come from a past so remote it is difficult for most people to grasp, our growing understanding of dinosaurs is part of our present, and something to build and develop into the future. That way we can be sure that we will encourage the dinosaur enthusiasts of future generations to find their own way into the dinosaurs' graveyard, and bring back some of the magic that they will find there.

GLOSSARY

Technical terms used frequently are defined briefly here. Others are defined in the text where they are used.

Ankylosaur: Armoured dinosaurs (Ornithischia).
Bone bed: An accumulation of separate fossil bones, which may be from one or more species.
Carnosaurs: A group containing the large carnivorous dinosaurs (Saurischia).
Ceratopsians: Horned dinosaurs (Ornithischia).
Champsosaurs: A group of crocodile-like reptiles. Not dinosaurs.
Coelurosaurs: A group of "hollow-tailed" small carnivorous dinosaurs (Saurischia).
Coprolite: Fossilized excrement.
Cretaceous: The last period of the Mesozoic Era.
Cryptozoology: The study of evidence for animals believed to exist but not yet known to science.
Cycad: A group of living and fossil plants having features resembling ferns and conifers.
Dromaeosaurs: A group of small carnivorous dinosaurs, noted for a large claw on the hind foot (Saurischia).
Ectothermic: Body temperature dependent on heat from external sources
Endothermic: Body temperature maintained by internal energy generation.
Fauna: All the animals of a particular region and period.
Gastroliths: Stones found in the digestive system of an animal and used, or presumed to have been used, for grinding up plant material.
Genus (pl. genera): A group of animals containing one or more species. The first word in a scientific name (e.g., *Tyrannosaurus*) is that of the genus.
Hadrosaur: Duck-billed dinosaur, including both crested and uncrested types.
Hoodoo: A natural pillar of rock, resulting from erosion.
Ichnogenus (pl. ichnogenera): A group of trace fossils containing one or more ichnospecies.
Jurassic: The middle period of the Mesozoic Era.
K-TEC: Workshops dealing with dinosaur extinction and related matters, short for Cretaceous-Tertiary Environmental Change.
Marsupial: Belonging to a primitive group of mammals whose young are reared in a pouch. Living marsupials include kangaroos and opossums; more primitive forms lived at the same time as the later dinosaurs.

Mesozoic: The era of middle life, lasting between approximately 245 and 65 million years ago. Includes the Triassic, Jurassic, and Cretaceous periods.
Microvertebrates: Small bones and teeth from all groups of vertebrates (including dinosaurs), typically collected by washing quantities of loose sediments.
Ornithischia: "Bird-hipped" dinosaurs, including duckbills and armoured and horned dinosaurs.
Ornithomimids: "Bird-mimics." A group of ostrich-like carnivorous dinosaurs (Saurischia).
Pachycephalosaurs: "Bonehead" dinosaurs (Ornithischia).
Pal(a)eo-: "Ancient." A prefix commonly combined with other words, as in palaeobotany, palaeoecology, palaeoclimatology, and palaeogeography.
Pal(a)eontologist: A scientist who studies ancient life. Main specialists are vertebrate and invertebrate palaeontologists.
Palynology: The study of recent and fossil pollen and spores.
Panel mount: A dinosaur exhibited in a panel, visible from one side only.
Pangaea: A former supercontinent composed of the roots of all the present continents united in one land mass.
Peigan: A nation of the Blackfoot Confederacy, historically living in the western prairies in what is now southern Alberta and the adjacent United States.
Plesiosaurs: A group of generally long-necked, long-tailed marine reptiles. Not dinosaurs.
Prosauropods: A group of primitive dinosaurs commonly seen as ancestral to the sauropods (Saurischia).
Pterosaurs: A group of flying reptiles. Not dinosaurs.
Raptor: A fierce predator. Used for large birds of prey and similar dinosaurs.
Saurian: Reptile. From the same root as "saurus" in many dinosaur names.
Saurischia: "Lizard-hipped" dinosaurs, including carnivores and sauropods.
Sauropod: Generally large dinosaurs that are long-necked and -tailed, and include such familiar dinosaurs as *Diplodocus* and *Apatosaurus* (Saurischia).
Scute: Bony plate embedded in the skin, as in crocodiles and ankylosaurs.
Theropod: A group of carnivorous dinosaurs, which to many palaeontologists now includes their descendants, the birds (Saurischia).
Triassic: The earliest period of the Mesozoic Era. Dinosaurs appeared during this period.
Troodonts: A group of small carnivorous dinosaurs, closely allied to birds.

SELECTED SOURCES

In preparing this book I consulted many printed sources and some archival ones and learned much that is otherwise undocumented from colleagues and Internet sources. There are some inaccuracies in the published record, and it would be too much to expect that all sources would agree on every detail. In cases of doubt, I tried to select the earliest or what seems to be the most authentic source. In a book for a general audience it is not possible or desirable to give the source of every piece of data, but for those interested in learning more about the subject, I have listed major general sources and then provided brief notes for each chapter directing readers to the main published and archival sources I used. Many of the most important general and older sources are listed in my book Dinosaur Hunters, *and are not listed again here unless they have extensive Canadian content.*

AMERICAN MUSEUM OF NATURAL HISTORY

(http://www.amnh.org)

Norell, Mark, Eugene Gaffney, and Lowell Dingus. 1995. *Discovering Dinosaurs in the American Museum of Natural History.* New York: Alfred A. Knopf. 248pp.

Preston, Douglas J. 1986. *Dinosaurs in the Attic. An Excursion into the American Museum of Natural History*. New York: Ballantine Books. 308pp.

AUDIOTAPES

Langston, Wann. Interview by W.A.S. Sarjeant.

Sternberg, Charles M. Interview for Provincial Archives of Alberta, transcript.

AUTOBIOGRAPHIES

Dawson, J.W. 1901. *Fifty Years of Work in Canada. Scientific and Educational.* Edited by Rankine Dawson. London and Edinburgh: Ballantyne, Hanson & Co. 306pp.

BIBLIOGRAPHIES

Danis, Jane. 1988. Bibliography of Vertebrate Palaeontology in Dinosaur Provincial Park. *Alberta. Studies in the Arts and Sciences* 1(1): 225–234.

Fox, Richard C. 1970. A Bibliography of Cretaceous and Tertiary Vertebrates from Western Canada. *Bulletin of Canadian Petroleum Geology* 18(2): 263–281.

Sarjeant, W.A.S. (comp.) 1980–96. *Geologists and the History of Geology. An International Bibliography from the Origins to 1993.* New York: Arno Press, and Malabar, Florida: Robert E. Krieger Publishing Company. 10 vols.

BIOGRAPHIES

Brown, Frances R. 1987. *Let's Call Him Barnum.* New York: Vantage Press. 81p.

Inglis, Alex. 1978. *Northern Vagabond. The Life and Career of J.B. Tyrrell — The Man who Conquered the Canadian North*. Toronto: McClelland & Stewart. 256pp.

Rogers, Katherine. 1991. *The Sternberg Fossil Hunters: A Dinosaur Dynasty*. Missoula, Montana: Mountain Press Publishing Company. 288pp.

Winslow-Spragge, Lois. 1993. *No Ordinary Man. George Mercer Dawson.* Edited by Bradley Lockner. Toronto: Natural Heritage/Natural History Inc. 208pp.

CANADIAN MUSEUM OF NATURE/NATIONAL MUSEUM OF CANADA

(http://nature.ca/)

Annual reports, bulletins, Syllogeus.

CANADIANS OUTSIDE CANADA

(see also Sino-Canadian Dinosaur Project)

Russell, Dale A. 1989. *An Odyssey in Time. The Dinosaurs of North America.* Toronto: University of Toronto Press, in association with the National Museum of Natural Sciences. 240pp.

Sarjeant, W.A.S. 1974. A History and Bibliography of the Study of fossil vertebrate footprints in the British Isles. *Palaeogeography, Palaeoclimatology, Palaeoecology* 16: 265–378.

CONFERENCES AND COMPILATIONS

Czerkas, Sylvia Massey, and Everett C. Olson (eds.). 1987. *Dinosaurs Past and Present.* California: Natural History Museum of Los Angeles County and University of Washington Press. 2 volumes. 149, 161 pp.

Folinsbee, R.E., and D.M. Ross. 1965. *Vertebrate Palaeontology in Alberta. Report of a conference held at the University of Alberta August 29 to September 3, 1963.* Edmonton: University of Alberta. 76pp.

Foster, John, and Dick Harrison (eds.). 1988. *Alberta. Studies in the Arts and Sciences 1(1).* Special issue on the Tyrrell Museum of Palaeontology. 246pp.

Russell, D.A., and G. Rice (eds.). 1982. *K-TEC II. Cretaceous-Tertiary Extinctions and Possible Terrestrial and Extraterrestrial Causes.* Ottawa: National Museum of Natural Sciences. Syllogeus No. 39. 151pp.

DINOSAUR GROUPS

Currie, Philip J. 1991. *The Flying Dinosaurs. The Illustrated Guide to the Evolution of Flight.* Red Deer, Alberta: Red Deer College Press: Discovery Books. 160pp.

Dodson, Peter. 1996. *The Horned Dinosaurs. A Natural History.* Ewing, N.J.: Princeton University Press. 346pp.

Feduccia, Alan. 1996. *The Origin and Evolution of Birds.* New Haven: Yale University Press. 420pp.

Horner, John R., and Don Lessem. 1993. *The Complete T. rex.* New York: Simon & Schuster. 238pp.

Paul, Gregory S. 1988. *Predatory Dinosaurs of the World. A Complete Illustrated Guide.* New York: Simon & Schuster. 464pp.

DINOSAUR PROVINCIAL PARK

(http://www.gov.ab.ca/~env/nrs/dinosaur/index.html)

Enns, Deborah, Robert Enns, Sandra Leckie, and John Walper. n.d., c. 1986. *Tyrrell Museum of Palaeontology and the Drumheller Valley*. Duchess, Alberta: Wildland Publishing. 32pp.

Reid, Gordon. 1986. *Dinosaur Provincial Park.* Erin, Ontario: The Boston Mills Press. 64pp.

Spalding, David, and Andrea Spalding, et al. 1985. *Dinosaur Provincial Park Learning Resources Manual.* Edmonton: Alberta Recreation and Parks. 237pp.

DINOSAURS IN CANADA

Gross, Renie. 1985. *Dinosaur Country. Unearthing the Badlands' Prehistoric Past.* Saskatoon: Western Producer Prairie Books. 128pp.

Horner, John R., and Edwin Dobb. 1997. *Dinosaur Lives.* New York: HarperCollins. 244pp.

Lambert, David and the Diagram Group. 1990. *The Dinosaur Data Book. Facts and Fictions about the World's Largest Creatures.* New York: Avon Books. 320pp.

Lessem, Don. 1992. *Kings of Creation.* New York: Simon & Schuster. 367pp.

Psihoyoos, Louie, and John Knoebber. 1994. *Hunting Dinosaurs.* New York: Random House. 267pp.

Reid, Monty. 1990. *The Last Great Dinosaurs.* Red Deer, Alberta: Red Deer College Press. 184pp.

Russell, Dale. A. 1977. *A Vanished World. The Dinosaurs of Western Canada.* Ottawa: National Museums of Canada, National Museum of Natural Sciences. Natural History Series No. 4. 142pp.

Thurston, Harry. 1994. *Dawning of the Dinosaurs.* Halifax: Nimbus Publishing, and the Nova Scotia Museum. 91pp.

Eastend Fossil Research Centre

(http://www.lights.com/scotty/)

Encyclopedias and Textbooks

Currie, Philip J., and Kevin Padian (eds.). 1997. *Encyclopedia of Dinosaurs.* San Diego, Cal: Academic Press. 869 pp.

Farlow, James O., and M.K. Brett-Surman (eds.). 1997. *The Complete Dinosaur.* Bloomington, IN: Indiana University Press. 752pp.

Fastovsky, David E., and David B. Weishampel. 1996. *The Evolution and Extinction of the Dinosaurs.* Cambridge: Cambridge University Press. 460pp.

Lessem, Don, and Donald Glut. 1993. *The Dinosaur Society's Dinosaur Encyclopedia.* New York: Random House. 533pp.

Expeditions and Fieldwork

Brown, B. 1919. Hunting Big Game of Other Days. A Boating Expedition in search of fossils in Alberta, Canada. *National Geographic Magazine* 35: 407–429. 25 ill.

Sternberg, Charles Hazelius. (1917, 1932) 1985. *Hunting Dinosaurs in the Bad Lands of the Red Deer River, Alberta, Canada.* Introduced and edited by David A.E. Spalding. Edmonton: NeWest Press. 235pp.

Weston, Thomas Chesmer. 1899. *Reminiscences among the Rocks in connection with the Geological Survey of Canada.* Toronto: for the author by Warwick Bros & Rutter. 328pp.

Extinction

Stanley, Steven M. 1987. *Extinction.* New York: Scientific American Library, Scientific American Books Ltd. 242pp.

Festschrifts

Churcher, C.S. (ed.). 1976. *Athlon. Essays on Palaeontology in Honour of Loris Shano Russell.* Toronto: Royal Ontario Museum, Publications in Life Sciences, Miscellaneous Publications. 286pp.

Field Guides

Russell, L.S., and C.S. Churcher. 1972. International Geological Congress. Field Excursion A21. *Vertebrate Paleontology, Cretaceous to Recent, Interior Plains, Canada.* Ottawa: XXIV International Geological Congress. 46pp.

Fundy Geological Museum

(www.nova-scotia.com/fundygeomuseum/)

Geological Survey of Canada

Zaslow, Morris. 1975. Reading the Rocks. *The Story of the Geological Survey of Canada 1842–1972.* Toronto: Macmillan of Canada. 599pp.

Summary reports of progress.

Geology and Palaeontology of Alberta

Nikiforuk, Andrew et al. 1994. *The Land Before Us. The Making of Ancient Alberta.* Red Deer, Alberta: The Royal Tyrrell Museum of Palaeontology, and Red Deer College Press. 96pp.

History of Dinosaur Discovery

Colbert, Edwin H. 1968. *Men and Dinosaurs. The Search in Field and Laboratory*. New York: E.P. Dutton & Co. 283pp.

Russell, L.S. 1966. *Dinosaur Hunting in Western Canada.* Toronto: Royal Ontario Museum Life Sciences Contribution 70. 37pp.

Spalding, David. 1993. *Dinosaur Hunters, 150 Years of Extraordinary Discoveries.* Toronto: Key Porter Books. 310pp.

Museums in Canada

(see also individual listings for major museums)

Key, Archie. 1973. *Beyond Four Walls. The Origins and Development of Canadian Museums.* Toronto: McClelland and Stewart. 384pp.

Periodicals

Alberta Report, *Canadian*, *Canadian Geographic*, *Dinosaur Report*, *Discovery*, *Earth*, *Earth Science History*, *Equinox*, *Geos*.

Royal Ontario Museum

(http://www.rom.on.ca)

Dickson, Lovat. 1986. *The Museum Makers. The Story of the Royal Ontario Museum.* Toronto: Royal Ontario Museum. 214pp.

University of Toronto, Geological Studies

Royal Saskatchewan Museum

(http://www.gov.sk.ca/govt/munigov/cult&rec/rsm/)

Royal Tyrrell Museum of Palaeontology

(http://tyrrell.magtech.ab.ca/home.html)

Royal Tyrrell Museum of Palaeontology. 1993. *Official Guide to the Royal Tyrrell Museum of Palaeontology.* Drumheller: The Royal Tyrrell Museum Cooperating Society. 66pp.

Dinogramme, fact sheets, occasional papers, *Trackways*.

Sino-Canadian Dinosaur Project

Currie, Philip J. (ed.). 1993. Results from the Sino–Canadian Dinosaur Project. *Canadian Journal of Earth Sciences* 30(10&11): iii–iv, 1997–2272

Currie, Philip J. (ed.). 1996. Results from the Sino–Canadian Dinosaur Project, Part 2. Canadian Journal of Earth Sciences 33(4): v–vi, 511–648.

Grady, Wayne. 1993. *The Dinosaur Project. The Story of the Greatest Dinosaur Expedition Ever Mounted.* Edmonton: Ex Terra Foundation, and Toronto: Macfarlane Walter & Ross. 261pp.

Strong-Watson, Linda, and John Acorn. 1993. *Official Souvenir Catalogue. Dinosaur World Tour*. Ex Terra Foundation in association with Macfarlane Walter & Ross. 96pp.

Video

Great North Productions. n.d. *In Search of the Dragon.*

Cinebus, National Film Board. 1978. *Charlie.*

Royal Tyrrell Museum Cooperating Society. 1993. *The Royal Tyrrell Museum Gallery Tour.*

MAJOR SOURCES BY CHAPTER

To save much repetition, reference to key names of individuals and in stitutions implies their published papers, biographical material (best accessed through Sarjeant's biographical index listed under Bibliographies on page 289), and histories of organizations.

CHAPTER 1

Huxley, Leonard. 1903. *Life and Letters of Thomas Henry Huxley.* London: Macmillan & Co. 3 vols. 463, 476, 501pp.

L'Heureux, Jean. Glenbow-Alberta Institute Archives.

Spalding, David. 1995. Bathygnathus, Canada's First "Dinosaur." In Sarjeant, W.A.S. (ed.). 1995. *Vertebrate Fossils and the Evolution of Scientific Concepts.* (no location given): Gorden and Breach Publishers. 245–254.

Thurston, Harry. 1990. *Tidal Life. A Natural History of the Bay of Fundy.* Camden East, Ontario: Camden House. 167pp.

Warkentin, John. 1964. *The Western Interior of Canada.* Toronto: McClelland & Stewart. 304 pp.

CHAPTER 2

Dawsons, Lambe, Tyrrell, Weston. American Museum of Natural History, Geological Survey of Canada.

Lambe, L.M. 1905. The Progress of Vertebrate Palaeontology in Canada. *Proceedings of the Royal Society of Canada* 2(10): 1:13–56.

Zeller, Suzanne. 1987. Inventing Canada. *Early Victorian Science and the Idea of a Transcontinental Nation*. Toronto: University of Toronto Press. 356pp.

CHAPTER 3

Brown, Cutler, Lambe, Sternbergs. American Museum of Natural History.

British Museum (Natural History). Sternberg letters and application to the Sladen Fund.

Charlie Sternberg audiotape.

Letters. Glenbow Archives, University of Alberta Archives.

CHAPTER 4

Brown, Cutler, Nopcsa, Parks, Sternbergs. American Museum of Natural History, Geological Survey of Canada, Royal Ontario Museum.

Colbert, Edwin H. 1992. *William Diller Matthew, Palaeontologist: The Splendid Drama Observed.* Irvington, N.Y.: Columbia University Press. 275 pp.

McAlpine, Donald F., et al. 1987. George F. Matthew . . . *New Brunswick Museum News* Aug–Sept. 26pp Natural History Museum, London.

CHAPTER 5

Brown, Parks, American Museum of Natural History, Calgary Zoo, Dinosaur Provincial Park, National Museum of Canada.

Letters. Glenbow Archives, and Natural History Museum, London.

Jones letters.

English, J. (ed.). 1979. The Calgary Zoo and Natural History Park. 1929–1979. *Dinny's Digest* IV(XI): 1–19.

Harle, Graham L. 1980. The Doctor and the Dinosaurs. *Heritage* 8(6): 18–19.

CHAPTER 6

Currie, Langston, Dale Russell, Loris Russell, Swinton, Dinosaur Provincial Park, National Museum of Canada, Provincial Museum of Alberta, Royal Ontario Museum, University of Alberta.

Folinsbee, R.E., and D.M. Ross. 1965. *Vertebrate Palaeontology in Alberta. Report of a conference held at the University of Alberta August 29 to September 3, 1963.* Edmonton: University of Alberta. 76pp.

Noble, Brian, and Glenn Rollans. 1981. *Alberta, The Badlands.* Edmonton: Reidmore Books. unpag.

CHAPTER 7

Currie, Horner. Dinosaur Provincial Park, Royal Tyrrell Museum.

Spalding, David A.E. Field notes, planning documents.

Peart, Bob. 1988. Face-to-Face: A Review of the Tyrrell Museum Exhibit Galleries. *Alberta. Studies in the Arts and Sciences* 1(1): 69–74.

Storer, J.E. 1986. An innovative look at Life's past: Tyrrell Museum of Paleontology. *Muse*: 51–54.

CHAPTER 8

Donald Baird, Currie, Dodson, Horner.

Dinosaur lists supplied by Philip Currie.

The "league tables" are in Lambert, David and the Diagram Group. 1990. *The Dinosaur Data Book. Facts and Fictions about the World's Largest Creatures.* New York: Avon Books. 320pp.

CHAPTER 9

Fox, Dale Russell, Royal Tyrrell Museum.

Interviews with Braman, Currie, Dennis, Eberth, Sloboda, and Sues.

Dixon, Dougal. 1988. *The New Dinosaurs. An alternative evolution.* New York: Fawcett Columbine. 120pp.

Russell, D.A., and R. Séguin. 1982. *Reconstructions of a small Cretaceous Theropod Stenonychosaurus inequalis and a Hypothetical Dinosauroid.* Ottawa: National Museum of Natural Sciences, National Museums of Canada. Syllogeus No. 37. 43pp.

Sarjeant, W.A.S. 1990. Astrogeological Events and Mass Extinctions: Global Crises or Scientific Chimerae? *Modern Geology* 15: 101–112.

CHAPTER 10

Currie, Reid, Dale Russell.

CHAPTER 11

Currie, Dale Russell, Sarjeant, Royal Tyrrell Museum.

Interview with Ebbels, Noble.

Cutler letters, Natural History Museum.

Grady, Wayne. 1993. *The Dinosaur Project. The Story of the Greatest Dinosaur Expedition ever mounted.* Edmonton: Ex Terra Foundation, and Toronto: Macfarlane Walter & Ross. 261pp.

Leakey, L.S.B. 1967. *White African.* New York: Ballantine Books. 274 pp.

Purvis, Andrew. 1998. Call Him Mr. Lucky. *Time* 151(26): 52–55.

Thurston, Harry. 1994. *Dawning of the Dinosaurs.* Halifax: Nimbus Publishing, and the Nova Scotia Museum. 91pp.

CHAPTER 12

Atwood, Cooley, Kish, Kroetsch.

Interview with Reid.

Atkinson-Keene, Susan. 1989. *Weekend in the Jurassic.* Halifax: Nimbus Publishing. 144pp.

Atwood, Margaret. 1979. *Life Before Man.* Toronto: Bantam-Seal. 292pp.

Blondin, George. 1997. *Yamoria: the Lawmaker. Stories of the Dene.* Edmonton: NeWest Press. 239pp.

Bullchild, Percy. 1985. *The Sun Came Down.* New York: Harper & Row, 390pp.

Kingsmill, Suzanne. 1990. Recreating the World of the Dinosaurs. *Canadian Geographic* 110(2): 16–27.

Kroetsch, Robert. 1975. *Badlands.* Toronto: New Press. 270pp.

Numbers, Ronald L. 1992. *The Creationists. The Evolution of Scientific Creationism.* New York: Knopf. 458pp.

Purdy, Al. 1984. *Piling Blood.* Toronto: McClelland & Stewart. 144pp.

Reid, Monty. 1993. *These Lawns.* Red Deer, Alberta: Red Deer College Press. 64pp.

Reid, Monty. 1985. *The Alternate Guide.* Red Deer, Alberta: Red Deer College Press. 64pp.

Richler, Mordecai. 1987. *Jacob Two-Two and the Dinosaur.* Toronto: McClelland & Stewart. 85pp.

Shuker, Karl P.N. *In Search of Prehistoric Survivors.* Blandford. 192pp.

Sawyer, Robert J. (1992–94) *Far Seer, The Fossil Hunter,* and *Foreigner.* New York: Ace Books.

Sawyer, Robert J. 1993. 'Just Like Old Times.' In Resnick and Greenberg (eds.). 1993. *Dinosaur Fantastic.* New York: Daw. 331 pp.

Twigg, Alan. 1981. *For Openers. Conversations with Canadian Writers.* Madeira Park, B.C.: Harbour Publishing. 271pp.

Wilson, John.1997. *Weet's Quest.* Toronto: Napoleon Publishing. 161 pp.

Wilson, John.1995. *Weet.* Toronto: Napoleon Publishing. 148 pp.

CHAPTER 13

Tanke, Tokaryk, Fundy Geological Museum, Royal Tyrrell Museum.

Interviews with Adams, Currie, Ebbels, Fedak, Naylor, Sues.

Environment Conservation Authority publications.

Alberta Heritage Act.

INDEX

T

U

V

W

Where to find Dinosaurs in Canada

The main areas of outcrop of Mesozoic rocks are indicated (though they are often covered with glacial deposits). Cities and towns with important dinosaur exhibits are marked, as well as a few key localities.

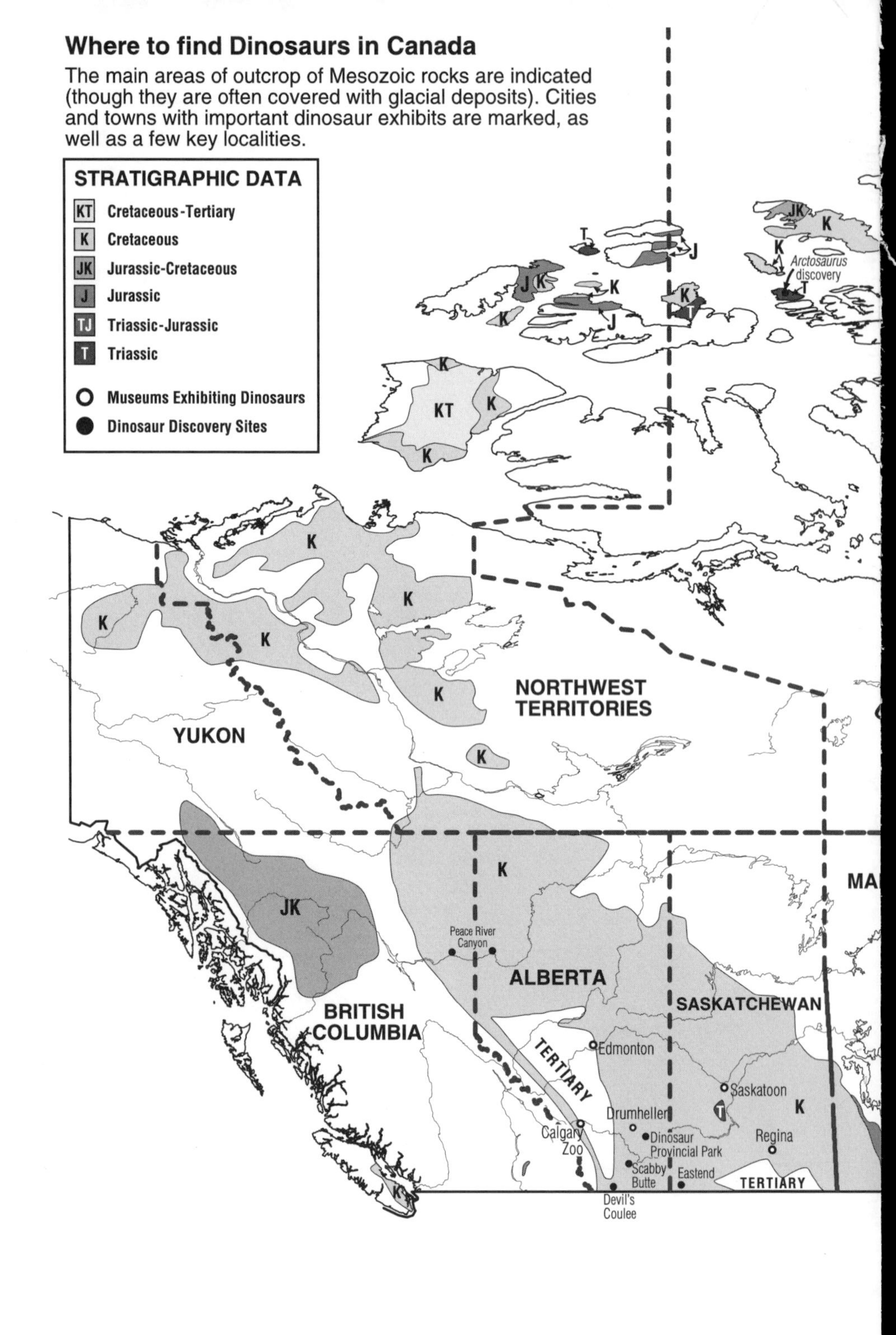